A Guide to

Patient Care Technology

A Review of Medical Equipment

For my family.
To quote Bryan Adams:

"Everything I do,
I do it for you."

A Guide to Patient Care Technology

A Review of Medical Equipment

Laurence J. Street BSc, Dipl Tech
Biomedical Engineering Technologist
Chilliwack General Hospital
Chilliwack, BC, Canada

The Parthenon Publishing Group
International Publishers in Medicine, Science & Technology

A CRC PRESS COMPANY
BOCA RATON LONDON NEW YORK WASHINGTON, D.C.

Published in the USA by
The Parthenon Publishing Group
345 Park Avenue South, 10th Floor
New York, NY 10010, USA

Published in the UK and Europe by
The Parthenon Publishing Group
23–25 Blades Court, Deodar Road
London SW15 2NU, UK

Library of Congress Cataloging-in-Publication Data
Data available on request

British Library Cataloguing in Publication Data
Data available on request

ISBN 1-84214-127-9

Printed and bound by Antony Rowe Ltd., Chippenham, Wiltshire, UK

Contents

PREFACE

A very wide variety of technological devices is used in the course of treating patients in clinical settings. This variety is increasing, and individual devices are often becoming more complex as well. The aim of this book is to examine the range of patient care technology that is commonly encountered in hospitals and clinics, and to give a clear and concise description of the principle of operation of each device, as well as an outline of the way it is used.

Medical caregivers using these instruments can be more effective if they understand the principles by which the machines work. They can use the equipment better, and they can solve simple operational problems more efficiently if they know how a device functions. These points apply to students in medical fields as well as current practitioners.

Patients and families reading this book will become informed about the equipment being used in their care. This will, hopefully, allow them to be more educated about their treatment, as well as making their stay in hospital somewhat less confusing and intimidating.

The table of contents is divided into sections corresponding to various areas of a hospital. Each section contains descriptions of equipment found primarily in that area. Since many devices are found throughout the hospital, a special section deals with them. At the beginning of each section, a listing is given of other equipment that might also be found the same area of the hospital.

Photographs of representative devices are included, though there may be considerable variation between different models. These photographs are mostly of devices 'in the field', rather than studio photos, to better give the reader a feel for their appearance while in use. All images, unless otherwise noted, are by the author.

After the description of each device, there are three short lists: other names by which the device might be known; equipment used with or related to the device; and locations where the device is most likely to be found.

It is not the intent of this book to replace user manuals. Devices performing the same function may be quite different in appearance and in details of operation from one manufacturer to another or even from one model to another, and the specific operational details must be obtained from the manufacturer-supplied user manuals. The information given here will supplement that which is given in the user manuals, and will provide a single source of information about a wide variety of equipment. Students can take the book with them for study, something often not possible with user manuals.

Following the equipment descriptions is an alphabetical index of devices featured in the book.

About the Author

Laurence J. Street earned a Diploma in Biomedical Engineering Technology at the British Columbia Institute of Technology in Vancouver, BC, Canada. He has since worked continuously in a variety of hospitals, from the very large Shaughnessey/B.C. Children's/Vancouver Women's complex to a regional hospital in the British Columbia Interior, and is employed presently in the Chilliwack General Hospital, a mid-sized community hospital outside Vancouver. In the latter two cases, he was solely responsible for setting up and implementing the Biomedical Engineering Departments.

Street received a BSc degree in Zoology from the University of British Columbia, and a British Columbia Teaching Certificate. He has taught Jr. High School science, as well as a number of electronics and math courses at college level, including several courses for electronics and computer technology that he developed himself.

Street's hospital work involves the repair and maintenance of all patient care electronic devices in the hospital. He is also closely involved in planning for future technological directions and the evaluation and acquisition of patient care equipment. He provides in-service education to medical staff regarding the safe and effective use of these patient care devices. Chilliwack General is a well-equipped acute care and teaching hospital, and is also part of a regional medical community, so he works with an extremely wide variety of both older and very modern equipment. He has taken a large number of factory training courses on the various devices that he services.

1. General

The types of medical devices described in this section are found throughout a typical hospital, in many different areas. Some of the devices such as stethoscopes are simple, while others, such as ECG machines or IV pumps, are complex and highly technical.

Alternating Pressure Mattresses

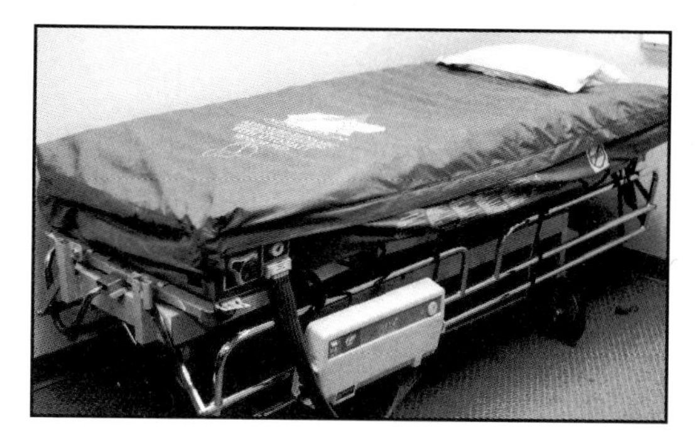

Overview

Patients with limited mobility are often subject to bed or pressure sores, as they are unable to shift their position to allow blood to flow to all parts that come in contact with the bed. While staff can provide changes in position, this isn't always practical, and during the night it might disturb the patient by repeatedly interrupting sleep.

Function

Alternating pressure mattresses have many separate sections, usually in two alternating groups, which can be inflated independently. By first inflating one group, then partially deflating it and inflating the second group, the mattress never is in firm contact with a given point on the patient for an extended period of time. The cycle time of the mattress is generally fixed at several minutes, as a compromise between comfort and effectiveness. Mattress pressure can be adjusted between preset limits to accommodate different patients and situations. Most systems can also be set to a non-alternating, or static, mode, in which all sections of the mattress remain inflated at the selected pressure.

The pumps for these systems have low and high pressure alarms, and the mattresses have valves that allow for rapid deflation in case this is necessary (for example, to allow cardiopulmonary resuscitation (CPR) to be performed). The valves can also be closed off completely so that the mattress remains inflated while the pump is removed or unplugged, such as when the patient is being moved to a different area. The low pressure alarm is inactive for several minutes after the pump is first turned on to allow for initial inflation.

Because the pumps are used for long periods close to the patient, they must be designed to be as quiet as possible.

Application

The patient is placed on the deflated mattress, which is then inflated by the pump. Pressure and mode are set once the mattress is fully inflated.

Also known as

Pillow pumps, pressure mattresses.

Related devices

Electrical hospital beds.

Where found

Areas of the hospital where patients are confined to bed for long periods, such as extended care, burn, or palliative care units.

Aspirators

Overview

In emergency situations, the patient's airway must be cleared of fluids that might block airflow.

Function

An aspirator is simply a suction pump that provides vacuum to a tubing system that connects to a specially-designed tip that is inserted into the patient's throat to remove fluids. The tip has openings that help prevent intake of larger objects, and are less likely to be blocked by surrounding tissue than would be a single, larger opening.

The glass or plastic reservoir between the pump and the patient serves two purposes: to collect the fluids drained from the patient; and to even out the air flow in the system by acting as a vacuum reservoir.

Other important components of the system include: a vacuum gauge to indicate the amount of vacuum available (and also give an indication that the system is set up correctly – a low reading may mean a leak); a vacuum level control mechanism, usually a knob; a filter system that helps to keep fluids from entering the pump and to prevent harmful materials from being sent into the air with the pump exhaust; and an overflow device to shut the system off if the reservoir becomes full of fluid. The overflow device can be a simple float valve, or a switch mechanism that depends of the weight of the reservoir to turn off the vacuum.

Pumping for the system can be accomplished by an electrical pump (either AC or battery-powered) directly associated with the rest of the components, or by a central pump that connects via a plumbing system to vacuum 'outlets' (of course they are actually 'inlets') at each bedside or treatment station. Self-powered systems are more portable, but central systems make for less clutter at the bedside.

Fluids collected by aspiration systems are non-sterile and must be treated appropriately.

Application

The pump is turned on and suction level adjusted, then the suction tip is applied to the area to be drained. Care must be taken not to damage tissues with excessive suction.

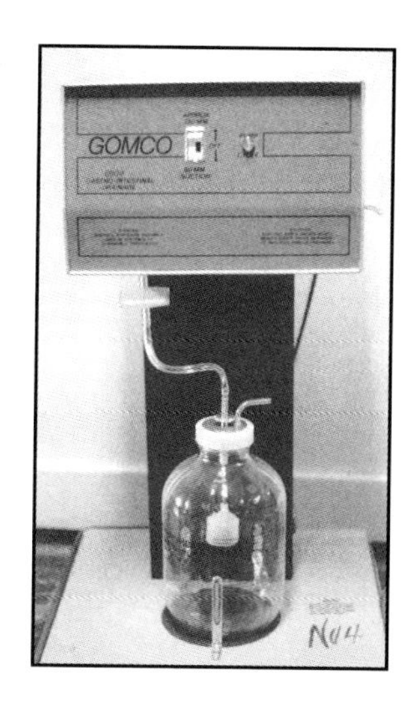

Also known as

Suction units, suctions, Gomcos (after a manufacturer).

Related devices

Wound drainage systems.

Where found

Most areas of the hospital.

Capnographs

Overview

In many medical situations, it is important to know how much carbon dioxide (CO_2) is present in the exhaled breath of the patient.

Function

A capnograph, which can be either a stand-alone unit or a module of a multi-function monitor, measures values by passing part of the exhaled air through a small chamber with glass sides. By passing precisely-controlled beams of infrared light through the chamber and measuring the output, the CO_2 content can be determined. The percentage is then displayed as a numeric value and/or as a graph on a display or recorder. If the capnograph is part of a monitoring system (either a module or a built-in component), the values may be stored over time to provide tabular or graphical displays of data which show changes over time.

Application

The capnography sensor is applied to the measurement window, which is usually in line with a ventilator or breathing tube. Because of the sensitivity of the measurements, the sensor must be calibrated before each use, and at specific intervals while in use. A pair of chambers is provided which have specific CO_2 content, and the measurements can be adjusted to match these values. A calibration factor, specific to that particular sensor, may be marked on the calibration portion of the sensor; if so, this value is entered into the system at a specific point in the calibration process. Once the sensor is calibrated, measurements can commence.

Also known as

CO_2 analyzers, exhaled CO_2 analyzer, end-tidal CO_2 monitors.

Related devices

Oxygen analyzers, point of care blood analyzers, ventilators, physiological monitors.

Where found

Critical care areas of the hospital.

Electrocardiograph (ECG) Machines

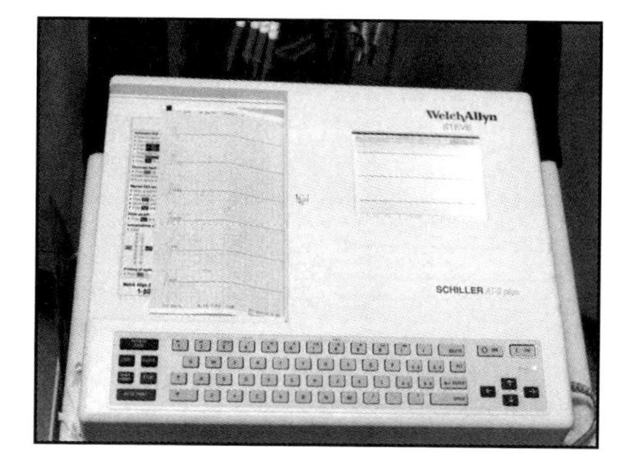

Overview

When a patient might have cardiac problems, caregivers need the most accurate indication of the functioning of the heart possible.

Function

One means of providing an indication of cardiac function is with the ECG machine, which makes precise measurements of the heart's electrical activity from several different perspectives, by using an array of electrodes (usually either five or twelve) placed in carefully-selected positions on the patient's chest. The circuitry of the machine then translates these signals into a graph and prints them out on chart paper, with one trace for each of several combinations of electrodes (or 'leads'). Since the signals are very small (measured in thousandths of a volt), compared to much larger electrical interference signals, the circuitry must be very sophisticated to provide useful results.

Most newer ECG machines provide annotations on the chart, giving lead labels, time and date, patient information, etc. Some provide precise values for various measurements of the ECG signal, and some also give a possible interpretation of any abnormalities present in the signals. The more complex machines usually have some kind of display screen and a keyboard for entering information. Some have data ports for connection to computer systems and/or to allow sending the results through a modem to a remote location.

Electrodes are usually either an adhesive patch with a conductive area in the center, or a silver cup with a rubber suction bulb that literally sucks the cup onto the patient's skin. The suction type are used with a conductive cream and are quicker and easier to apply and remove; adhesive electrodes are less obtrusive and stay in place better for longer term applications.

Most ECG machines have batteries that allow them to be used when access to a power outlet is difficult; many also have data storage so that records can be printed out at a later time, since they are often used in the midst of an emergency situation and need to be removed from the scene as quickly as possible.

Application

The patient's skin is prepared for electrodes by shaving the sites (if necessary) and then cleaning the surface thoroughly. Natural or artificial oils on the skin can reduce contact quality, but the salts in the skin increase conductivity, so gentle physical abrasion of the skin is preferred over soap and water or alcohol cleaning. Electrodes are then applied; accuracy of placement in the prescribed locations is critical for obtaining the best results. Most systems allow the operator to view the signal before making a recording; when the setup is complete, a recording is initiated. This usually only takes a few minutes, though some situations call for recordings made over a longer period of time.

Also known as

Electrocardiographs, EKG machines (from the German spelling of the term), EKGs, ECGs, 12 leads, ECG/EKG recorders.

Related devices

ECG monitors, Physiological monitors, Ambulatory ECG recorders, Stress test machines.

Where found

Units are usually mobile and can be taken to any area of the hospital where patients might have cardiac problems, but are usually seen in emergency rooms, intensive care units, post anaesthesia recovery areas, or cardiology rooms.

Electric Hospital Beds

Overview

Hospital patients need to be able to change their position in order to be as comfortable as possible, especially if they are in bed for long periods of time. It is useful if the patient can adjust the bed to become more comfortable; it is also valuable to staff to have the patient make position changes unassisted.

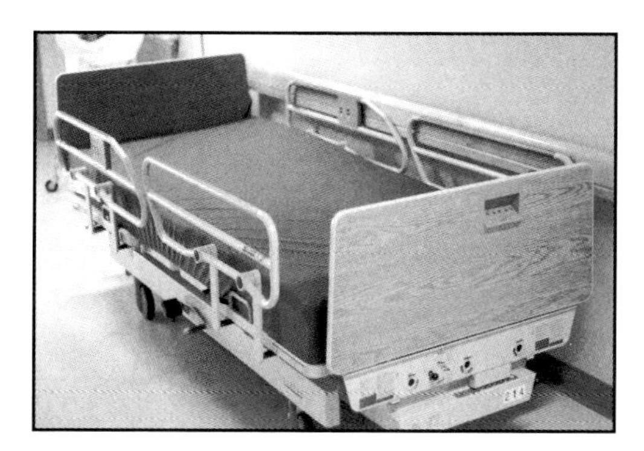

Function

Electric hospital beds have a comfortable, flexible mattress, and a frame and mechanism to allow for a variety of patient positions. Controls activate motors that raise or lower the whole bed, the head, the feet, and the hips independently (within ergonomic limits); these controls are usually placed both on a pendant or stalk so that the patient or a bedside attendant can use them, and at the foot of the bed for attendant control. Sometimes duplicate controls are on the bed rails on either side of the mattress. There is usually a lockout switch so that caregivers can prevent the patient from activating the controls, if this is necessary.

The whole unit is on wheels, which can be locked to prevent movement when necessary. Rails on either side of the bed can be raised or lowered, often in multiple positions, to help prevent the patient from falling out of bed. As with all hospital beds, the mattress is covered with an impermeable material to aid in cleaning.

Application

The bed is adjusted to suit the patient's comfort and/or treatment situation.

Also known as

Power beds.

Related devices

Birthing beds.

Where found

General patient wards.

Electronic Probe Thermometers

Overview

Body temperature is a critical vital sign.

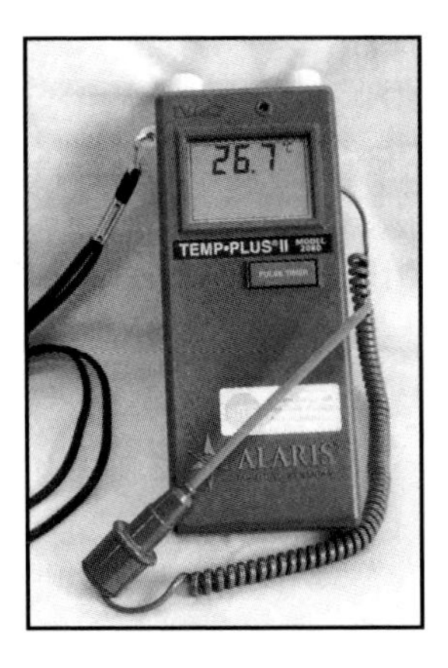

Function

Probe thermometers use a tiny electronic device to measure the temperature of the body part to which they are applied. The temperature is then displayed on a digital read-out, which may be on a small, hand-held box or on the screen of a larger, multi-function monitor. Hand-held units usually have the sensor on the end of a thin rod, which can be placed under the tongue, under the arm, or rectally. To ensure patient safety, there are separate, color-coded probes for oral and rectal use, and the probes are covered with a smooth, tough disposable cover which is changed for each patient. Often, there is a timer in the hand-held part which beeps when the temperature reading has had time to stabilize, or the unit can be set to continuously measure the patient's temperature. For longer-term monitoring, sensors on a small, flexible wire are available; these can be placed on the patient and not interfere too much with movement.

Units usually have a switch to select between Celsius (Centigrade) and Fahrenheit degrees.

Application

A clean cover is placed on the probe, and it is applied to the patient. After an appropriate time, the temperature reading is recorded, the probe is removed, and the cover discarded.

Also known as

Electronic thermometers, digital thermometers, probe thermometers.

Related devices

Tympanic thermometers, physiological monitors.

Where found

Most patient care areas of the hospital, but because most intensive care unit (ICU) and operating room (OR) multifunction monitors have built-in thermometers, they may not be used in these areas.

Defibrillators

Overview

The contraction of the heart is controlled by electrical signals, which must be of the correct size, timing, and distribution to cause coordinated and effective pumping. Many factors can disrupt these signals, such as blood electrolyte and/or gas levels, body temperature, various drugs, damage to the heart muscle or to the conductive pathways that distribute the signals, and physical or electrical shock. When the heart no longer beats effectively, blood flow is compromised, and serious harm or death can result. Note that the heart may continue to beat in some fashion, but not effectively (that is, it fibrillates or is in fibrillation), or the heart may cease to beat at all (asystole, 'flat line'). Whatever the cause, it is essential that proper function be restored as soon as possible. This may be accomplished by medication, by physical intervention such as CPR or body temperature adjustment, or by applying an electrical signal to the heart that will restart or re-coordinate (defibrillate) its contractions. A defibrillator is a device that supplies such corrective signals.

Function

Defibrillators may act on the heart directly, with electrodes applied to the cardiac tissue, or indirectly, with electrodes (called paddles) placed on the exterior of the patient's chest. The corrective signal, or shock, required is much, much greater when the electrodes are external, since a large portion of the signal will be blocked or redirected by the tissues between the electrodes and the heart itself. In most emergency situations, it is not possible to apply the electrodes directly to the heart, so most defibrillators are of the external type.

Internal defibrillators can be either implantable, in which case they must be capable of determining when shocks are necessary and administering them automatically, or manual, in which case the size and timing of the shock is determined by the operator. Most manual internal defibrillators are simply regular external defibrillators with specially-designed spoon-like electrodes for direct application to the heart (for example, during open-heart surgery). These defibrillators must be capable of delivering the small-sized shocks required, and most have a circuit that detects when the internal paddles are being used to prevent the selection of higher shock levels.

Implantable defibrillators use wire electrodes that are embedded in the cardiac muscle at the appropriate location. They measure the ECG signal and analyze it to determine its characteristics. If these characteristics fall into a class that has been designated as abnormal but correctable, they will automatically administer an appropriately-sized-and-timed shock to re-establish proper cardiac function. This is different from implantable pacemakers, which apply smaller signals in order to maintain proper rhythms. Generally, a pacemaker is used when the degree of damage to the heart that requires corrective action is less, whereas an implantable defibrillator is used when there is greater damage and more extensive intervention is required to restore proper function.

External defibrillators may also be automatic; this is usually a feature of a machine that can also be used in manual mode.

Manually-controlled external defibrillators have some means of detecting and displaying and/or analyzing the ECG signals from the heart. This may be done directly through the paddles, or through separate ECG electrodes. Using separate electrodes generally produces more accurate ECG signals, but it takes time to apply them, and so the paddles themselves are used in more urgent situations. When separate electrodes are used, they must be located so as not to interfere with the optimum placement of the paddles. If the paddles are placed on top of the ECG electrodes, the shock can be reduced in effectiveness and/or burns can result to the patient's skin. Units often have a visual indicator of paddle contact quality.

Controls for the defibrillator include energy selection level, charge initiation, and discharge. The discharge buttons are usually a pair, one on each paddle handle, which must both be depressed to cause a discharge. Charge initiation and sometimes even energy selection controls may also be located on the paddles, as well as on the front panel.

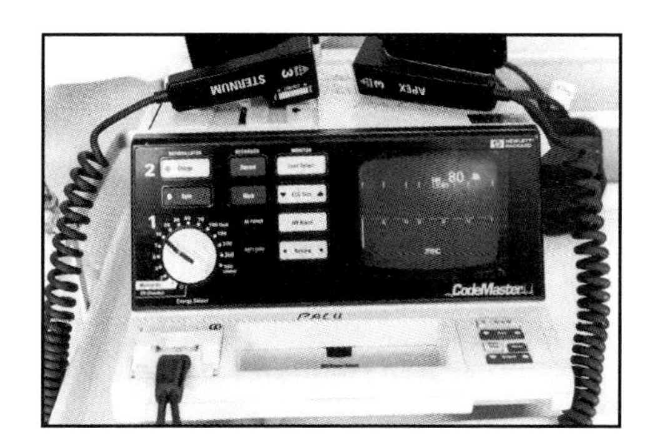

These units have some kind of display of the ECG signal. This may be a simple visual or audible indicator of each beat, but usually it is a graphic display of the ECG waveform. Along with the graphic display, there may also be an audible beat indicator, whose volume can be adjusted or turned off. There is often a paper recorder that prints the signal and usually indicates the point at which the shock is administered and its size, as well as the date and time, which can be important for later medical analysis or legal considerations. The recorder often will have a memory function which allows it to print out several seconds of ECG signal from before the time the shock was applied, which gives a better picture of the event as a whole.

Defibrillation may be required to correct other abnormalities of the cardiac rhythm. The most common of these is when the atria of the heart are fibrillating but the ventricles are beating normally (atrial fibrillation or afib). In this case, the shock must be applied at a very precise time in relation to the ventricular beats. Since this would be almost impossible for a human operator to time correctly, defibrillators are often equipped with circuitry to detect the ventricular beats and apply the shock at the next correct time after the operator depresses the shock (or discharge) buttons. This procedure is called synchronized cardioversion. Since the proper timing requires an accurate ECG signal, the defibrillator will not enter this mode unless ECG electrodes are applied and used. The mode is normally called 'sync' (synchronized), and there is usually a light flashing and/or tone sounding to indicate its selection. There is usually a 'marker' signal placed on top of the normal contraction signals so that the operator can determine whether the timing is correct.

Because they are often used in emergency situations, defibrillators often have internal batteries that can provide power for all functions.

Defibrillators may have other functions built in, such as pulse oximetry (SpO_2), in order to help caregivers better assess the condition of the patient.

Application

Once the need for defibrillation has been established, the patient must be at least minimally prepared. The patient must be placed in a safe position, ensuring that the defibrillator paddles can make good contact with the correct areas of the chest wall. Conductive materials such as saline, blood, and ECG electrodes must be clear of the site. The team leader then decides on the optimum energy setting (though it may be selected by another team member). If a synchronized cardioversion is to be performed, that function must be selected. A button is pressed to start the unit charging, and a tone often sounds to indicate that this is happening. Another tone signals that charging is complete, and then (usually after notifying the other team members first) the operator presses the discharge buttons, which causes the shock signal to be delivered to the patient. The ECG signal of the patient is then observed to determine whether the treatment was successful. If not, the energy level may be increased and the procedure repeated.

Because the shock delivered by a defibrillator is quite large, and because the signal may travel to other parts of the body than the heart, where it may cause sharp contractions of muscles, the patient may twitch rather violently when the shock is delivered. This means that appropriate measures must be taken to prevent injury to the patient or to attending staff.

Also known as

Defibs.

Related devices

Pacemakers, pulse oximeters, ECG monitors.

Where found

Critical care areas, general areas of the hospital on 'crash carts'.

Examination Lamps

Overview

In evaluating the condition of a patient, it is important to be able to see clearly.

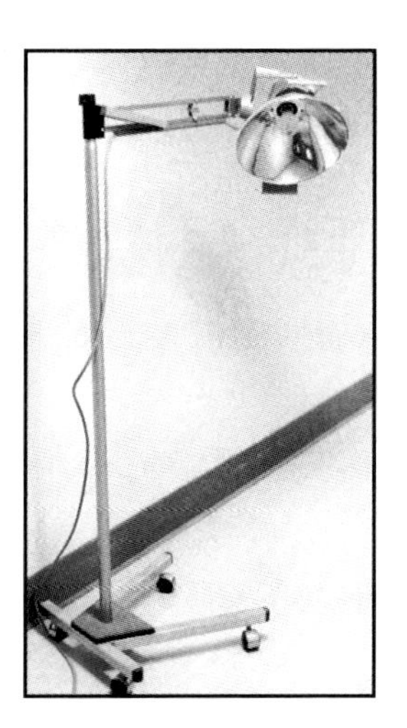

Function

Proper observation requires a light source that can easily be placed in a variety of positions, and which provides bright illumination, with no coloration that might obscure clinically significant details such as tissue color. Since the examination lamp may be used for extended periods, it shouldn't give off excessive amounts of heat. They may be ceiling- or wall-mounted, or on a moveable stand. Most are line powered, though some smaller units have batteries. Some have intensity adjustments.

Application

The light is turned on and intensity is set, then the unit is positioned to give the best illumination.

Also known as

Lamps, lights.

Related devices

Oto-laryngo-ophthalmoscopes, OR lights.

Where found

Emergency rooms, outpatient areas, examination rooms.

Feeding Pumps

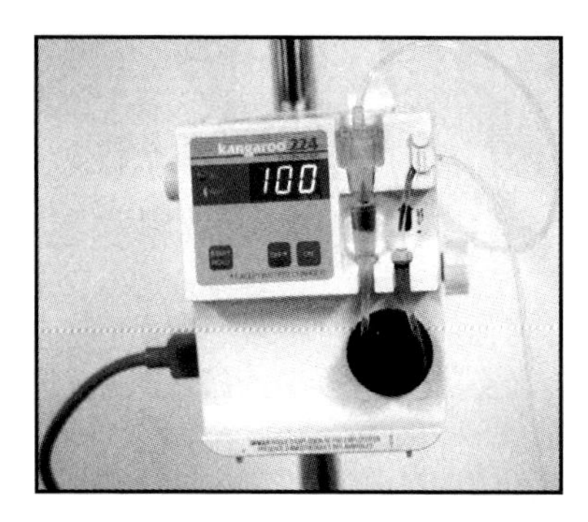

Overview

For a variety of reasons, a patient may not be able to swallow food adequately. In such cases, food must be administered through a tube inserted into the stomach. Obviously, the food has to be liquid, and as such liquids are generally quite thick, they may not flow easily enough to pass through the tube by gravity alone.

Function

Feeding pumps provide pressure to make the liquid food flow through the feeding tube and into the patient's stomach. The most common mechanism for these pumps is a wheel, around part of which an elastic section of tubing is stretched. Cylindrical roller bearings at intervals around the wheel compress the tubing as the wheel rotates, squeezing the fluid in the tube in a peristaltic wave. This pushes the liquid food through that section of tubing and toward the patient, while at the same time drawing more food from a reservoir. Pumps generally have controls that allow flow rate and desired volume to be set, and have some method of signaling when the set volume is reached. They often have a means of detecting blockages, and alarms for when this occurs.

Application

Use of feeding pumps is relatively straightforward. Once the feeding tube is inserted into the patient's stomach (usually through one nostril and down the esophagus, but sometimes through an incision in the abdomen and stomach wall), the elastic portion of the tubing is stretched around the pumping wheel and connected to the food reservoir. Flow rate and required volume is set, and the pump is started.

Also known as

Enteral feeding pumps, Kangaroo pumps (after one large manufacturer's model name).

Related devices

Intravenous pumps.

Where found

Most areas of the hospital, extended or continuing care units.

Gas Regulators

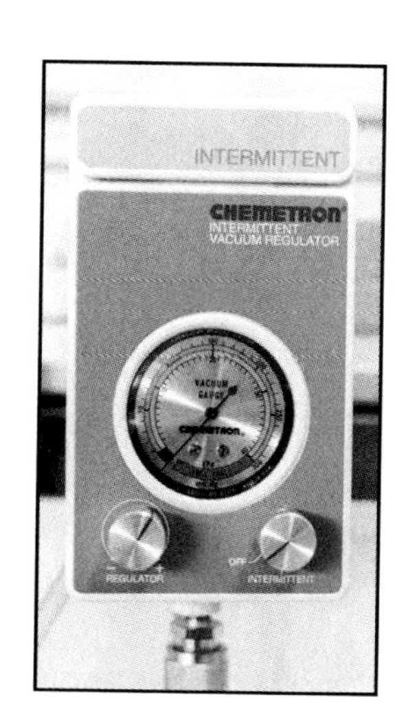

Overview

A variety of gases are used in routine patient care, particularly air (both supplied as compressed air and removed as vacuum) and oxygen. Some areas also might have nitrous oxide available.

In all cases, the gases are supplied by a source with a certain and relatively-high pressure. The source may be local, such as a smaller pressurized cylinder or portable suction pump, or remote, from much larger pressurized tanks or a central vacuum pump. This pressure is usually too great to apply directly to patients, so some kind of gas regulator must be used.

Function

Gas regulators are usually mechanical units that employ bellows and/or needle valves to control pressure and flow, and usually they have indicators for both of these parameters. The indicators may be dial gauges or digital displays. For flow readings, a small ball may be used in a tapered, marked tube; the gas flows up the tube and the ball rises to a height proportional to the flow rate. Units have controls (usually knobs) to control the flow and/or pressure, and may have alarms to indicate high or low pressures.

Application

Units are used to adjust the flow rate and pressure required for the specific situation.

Also known as

Oxygen (O_2), air, vacuum, suction, flow or pressure regulators/gauges/meters.

Related devices

Oxygen analyzers, aspirators.

Where found

All patient care areas.

Glucometers

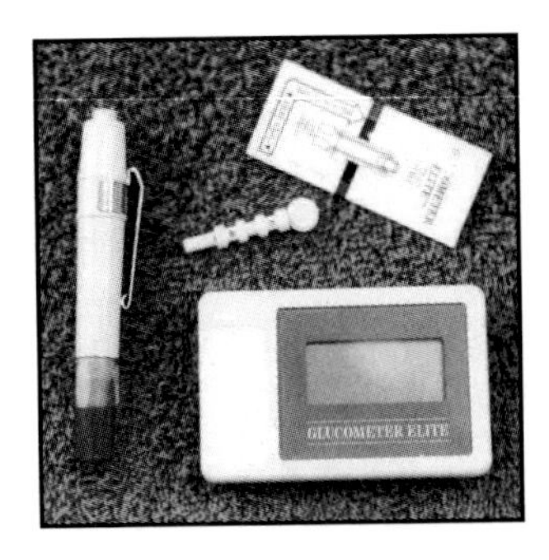

Overview

For patients with diabetes, and sometimes in other situations, it is important to be able to determine blood glucose levels.

Function

Devices for measuring blood glucose are small, reliable, accurate, and simple enough to be used by patients themselves, though of course medical caregivers use them as well. The devices rely on a chemical reaction in which glucose in the blood causes a color change on a small paper or plastic strip. A small drop of blood is applied to the strip, and after a pre-set time (usually signaled by the testing device), the strip is inserted into a port in the unit. With some models the strip is inserted and then the blood is added. Electronic circuitry measures the color change of the strip and from this data, determines the blood glucose level and displays it on a screen.

Because of the critical nature of this test, glucometers must be very accurate and stable. They have means by which the user compensates the unit to each batch of test strips, to account for slight variations from batch to batch. They also have calibration strips, which have precisely controlled colors, to check for proper operation.

Blood glucose levels can change quickly following meals or exercise, and measurements must be made with this in mind. Factors such as timing and type of food or drink taken and patient activity level must be noted, along with the reported blood glucose values.

Application

The unit is prepared and calibrated, and then a small blood sample is taken, usually from a lancet prick on a finger. The blood drop is placed on a strip from the same batch as the calibration strip, and the unit timer started. When the time is up, the strip is placed in the measuring slot, and the blood glucose level is recorded. Alternatively, the blood is added to an already-inserted strip.

Also known as

Blood sugar analyzers.

Related devices

Point of care blood analysis systems.

Where found

Most areas of the hospital, outpatient clinics, patients' homes.

Intravenous Pumps

Overview

Medications, blood, or fluid often must be delivered into a patient's bloodstream at an exact rate for a relatively long period of time. In the old days, caregivers had to measure how fast the liquid was dripping and calculate the rate, adjusting the flow often. The intravenous pump makes this much easier by pumping the fluid in a controlled and accurate manner.

Function

There are several different techniques of pumping. One of the most common is 'peristaltic', in which either a set of 'fingers' or some rollers on a drum squeeze the fluid tubing, one after another, pushing a wave of fluid through the tube. Because the tubing is precisely manufactured, the speed at which the pump operates determines the flow rate of the fluid. The other common pumping method uses a cassette that fits into the pump and is compressed by a piston. The size of the pumping chamber in the cassette is very precise, and, by varying the frequency of compression, the flow rate can be controlled. This method is generally more accurate than the peristaltic method. With either method, an internal computer calculates and displays the flow rate, and also the total volume to be infused, how much has been infused, and how much is left.

Pumps have very accurate sensors which can detect air bubbles in the line and stop the pump, sounding an alarm before the bubbles can get past the pump and into the blood stream. They also can detect blockages, either before or after the pump.

Application

After establishing an intravenous line, the tubing or cassette is carefully inserted into the pump. Infusion rate and total volume to be infused are set, and the pump is started. Some pumps can be programmed to pump at one rate for a set time or volume, and then change to another rate. For example, when two different doses of different medications are being given at once, the smaller dose is pumped first, then the other. Some pumps can control two or more different fluid lines at the same time.

Most pumps have an internal battery that allows all functions to continue when the pump is unplugged from the AC wall outlet, allowing the patient to go the washroom or to be moved to other areas or even a different hospital without interrupting pumping.

When the pump has delivered the programmed volume, an alarm sounds and a visual indicator is displayed. The alarm function may be interfaced with a nurse call system to alert personnel who may not be in the immediate vicinity. Usually, the pump continues to deliver a very small amount of fluid in order to prevent blood from clotting over the tip of the intravenous needle or catheter; this is referred to as a 'keep vein open' or KVO mode, and is usually accompanied by an intermittent tone.

Because tubing or cassettes may wear and/or become deformed with continued use, possibly affecting flow accuracy, they must be replaced regularly, in accordance with manufacturers' recommendations.

Also known as

IV pumps, IVACs (after one of the first manufacturers of such pumps), infusion pumps.

Related devices

Patient-controlled analgesia pumps, syringe pumps, feeding pumps.

Where found

Most areas of the hospital.

Invasive Pressure Monitors

Overview

It is often important to know pressure values within the patient's body. This is most commonly the pressure in limb arteries, but can also include blood pressures in the venous system, at various points of the arterial system close to the heart, or cerebrospinal fluid pressures in the skull or spinal column. Blood pressures are pulsatile, and the maximum (systolic), minimum (diastolic), and mean pressure values are also important.

Blood pressure can be measured at the limbs non-invasively (by NIBPs, see next entry), but this method is susceptible to errors, it is not continuous, and it cannot give values for anything but limb arterial pressure.

Function

By introducing a catheter with an electronic pressure transducer at its tip, exact measurements of pressure can be made continuously, at whatever point the catheter is positioned. Transducers are small and can be guided from a convenient insertion point to various locations within the circulatory system, either by simply watching the pressure values and knowing the typical measurements from various locations, or by viewing the catheter and vessels with x-rays.

Because of the continuous nature of these pressure measurements, a waveform of the values can be displayed on an associated monitor; the shape of the wave signal can be significant as well as the various pressure values. Systolic, diastolic, and mean values can be displayed; many monitors also record measurements over time so they can be displayed in tabular or graphic form. High and low alarm levels can be set, sometimes for systolic, diastolic, and mean values independently.

Pressure transducers are sensitive and must be calibrated prior to each use; they must also be 'zeroed' regularly by exposing them to atmospheric pressure in order to compensate for slight variations.

Application

A catheter with pressure transducer is established in the vessel or chamber to be monitored. The system is zeroed and calibrated as necessary, and measurements begun.

Also known as

Pressure monitors, BP monitors.

Related devices

Physiological monitors, NIBPs, cardiac catheterization units.

Where found

Critical care areas, operating and post-op recovery rooms, some special units (e.g. cath labs).

Non-invasive Blood Pressure (NIBP) Monitors

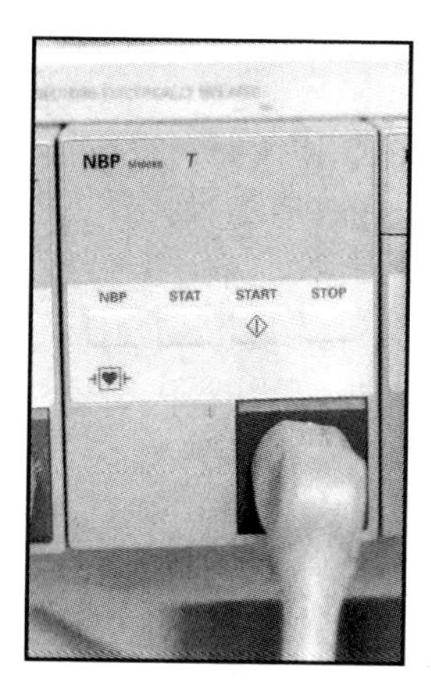

Overview

A patient's blood pressure is a vital measurement for diagnosis. When absolute accuracy is needed, and/or monitoring will be over an extended period, and/or when continuous instantaneous measurements are required, a catheter is inserted into the patient's blood vessels and pressure is measured directly by an attached transducer. This invasive technique is often not desirable, however, and is generally impractical for taking blood pressure readings manually at frequent intervals over an extended period of time.

Function

Non-invasive blood pressure monitors (NIBPs) allow blood pressure measurements to be made automatically. Most NIBP systems use a technique similar to that of manual blood pressure measurement. An inflatable cuff is placed over the patient's limb (usually the arm but sometimes the thigh). The cuff is inflated until arterial blood flow is occluded and then pressure is slowly released while monitoring the return of partial and then complete blood flow. The cuff pressure at which partial flow commences is equivalent to the systolic blood pressure, while the point at which complete flow is restored is equivalent to the diastolic pressure.

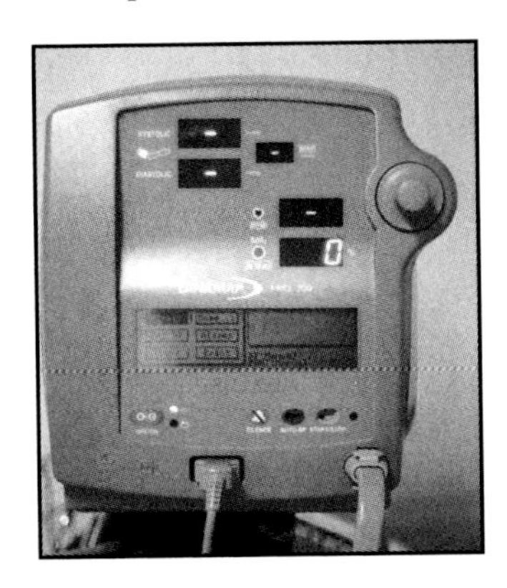

Two basic types of NIBPs are in common use. In one, a microphone placed near the patient's artery, usually under the inflatable cuff, determines the presence or absence of blood flow. Associated electronic circuitry processes the sound signals and analyzes the resulting waveforms to determine the systolic and diastolic points.

The second type of NIBP measures small variations in air pressure within the cuff caused by blood pulsing within the patient's arm. The pattern of these variations is used to determine blood pressure measurements.

In both systems, readings may be initiated by the operator, or may occur at regular intervals, the timing of which is determined by the operator. There is some kind of display for results; systolic and diastolic values may be displayed alternately in a single window, or they may each have separate displays. Mean pressure may be calculated and displayed as well.

Some systems have a 'stat' button, which causes the machine to take several measurements in quick succession. Many also have programmable alarm limits to notify staff when pressures are too high or too low. Another common feature is a recorder module, which allows printouts of results, along with the time they were taken, and sometimes a graph of pressure trends over time.

A new technology uses motion of the blood vessel wall to determine pressures, and will allow continuous instantaneous blood pressure measurements, without requiring the insertion of a catheter.

NIBPs may be stand-alone; they may be built into devices that also measure ECG and/or SpO_2, or they may be modules in a general physiological monitoring system.

Application

An appropriately-sized cuff is selected and applied to the patient's limb. Measurement interval is selected, and the unit is started. A blood pressure measurement is taken and the values are displayed. Measurements are repeated automatically at the selected intervals. Cuff size and placement are important for accurate readings.

Note that, because of the nature of the measurement process, values obtained by NIBP units may be different than those from manual or invasive measurement. Pressure values do not necessarily correspond exactly to absolute values within the blood vessels; however, they are accurate enough for most purposes and give a very good indication of trends in measurements over time.

Also known as

BP machines, Dinamaps (after one of the common early models), NBPs.

Related devices

Physiological monitors, invasive pressure units, sphygmomanometers.

Where found

Most areas of the hospital.

Oto/Laryngo/Ophthalmoscopes

Overview

The ear (oto), throat (laryngo), and eye (ophthalmo) are body parts that often require close examination. To do so effectively, equipment must provide bright, adjustable lighting that can illuminate the subject while allowing a clear view.

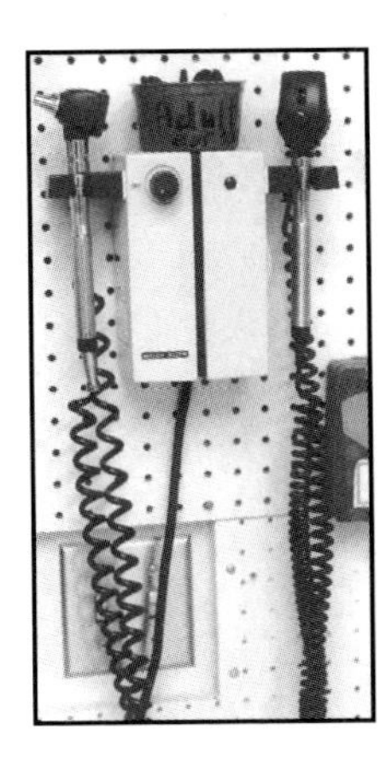

Function

Otoscopes typically have a tapered tube that can be inserted into the ear canal, with a magnifying lens (sometimes interchangeable for various magnifications) over the large end. Light is supplied by a small bulb in a handle that is either battery- or AC-powered. The light in most otoscopes is channelled from the bulb into the ear by a set of optical fibers embedded within the viewing tube, so that the light emerges in a ring around the end of the tube, thus providing light without interfering with viewing. Some otoscopes use a simpler arrangement, with a bulb shining directly down the tube and blocked from view by a shield. This shield, of course, obstructs the field of view somewhat. Disposable covers are usually used to help prevent the spread of infection from one patient to another.

Laryngoscopes need a means of holding the soft tissue in the patient's mouth (mainly the tongue) out of the way for viewing. They are usually equipped with a curved metal blade for this purpose. The blade has rounded edges and is made of metal so that it won't break if the patient inadvertently bites down. A bright light (either a bulb or the end of an illuminated fiber optic channel) is placed at the base of the blade to illuminate the area being viewed. Laryngoscope blades must be sterilized between patients.

Ophthalmoscopes have a bright light source as well, but since structures of concern in the eye are generally smaller than those in the ear or throat, ophthalmoscopes have lenses of various powers to aid in observations. Some eye conditions are visualized better with colored light, so filters are often available for the observation light. In addition, some situations require a narrow line of light to illuminate the eye, so mechanisms are provided to make such lines, usually of various widths.

OLO scopes can be powered with rechargeable or disposable batteries, or they may be connected to a power supply fed from a wall outlet. Since they are required for many emergency cases, the units in the ER are often wall-mounted and line-powered, so that dead batteries are not a problem.

Application

Scopes are used to examine the body part in question.

Also known as

Ear/throat/eye scopes.

Related devices

Slit lamps.

Where found

Most areas of the hospital

Oxygen Concentrators

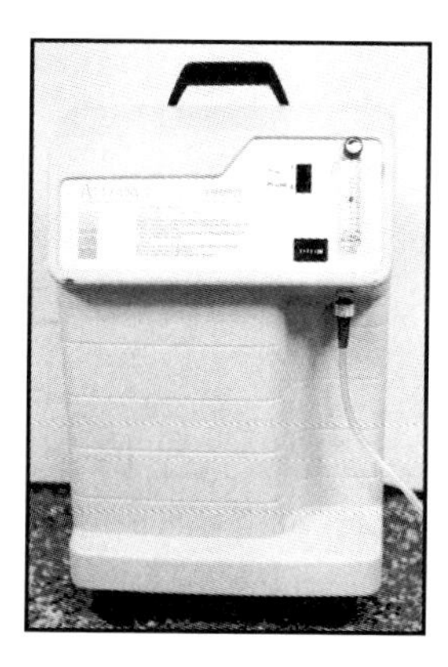

Overview

Some patients may be located in areas where there is no wall outlet to supply oxygen.

Function

In certain situations, an oxygen concentrator must be used. These devices utilize the fact that special resin bead materials absorb nitrogen; since nitrogen composes about 79% of air, and oxygen almost all of the remaining 21%, if the nitrogen can be removed, almost pure oxygen will remain.

Obviously, no material can absorb an infinite amount of anything, so oxygen concentrators use two sets of resin beads, placed in canisters called sieve beds. A system of valves, hoses, and pumps causes room air to flow through one sieve bed, which removes the nitrogen. The resulting oxygen is then delivered to the patient, at a rate set by the caregiver. When the first sieve bed is near to maximum capacity, flow is switched to the second sieve bed. Then air is flushed in a reverse direction through the first sieve bed, which removes the built-up nitrogen and vents it harmlessly back into the atmosphere. When the second bed is full, the system switches back to the first bed and the process starts again while the second bed is back-flushed. A small pressure tank near the outlet of the system maintains constant flow to the patient during switchovers.

Because these units operate for long periods of time and are providing vital oxygen, they must be very reliable. They will typically operate for several thousand hours before requiring service. Another design consideration, since they are used at the patient's bedside 24 hours a days, is that they be quiet, and considerable effort is put into silencing the mechanisms as much as possible.

Oxygen concentrators must have a warning alarm in the event of the flow of gas becoming restricted or if other potentially harmful conditions arise. Some have a built-in oxygen analyzer that measures the output oxygen concentration to ensure it is adequate, though most units require that an external analyzer be used on a scheduled basis to test for this. Generally, if the flow rate is maintained, the system is operating properly and oxygen levels are adequate. If the sieve beds are reduced in efficiency, flow decreases and the flow alarm sounds.

Application

The unit is turned on, and, after a few cycles, the oxygen flow rate is adjusted to meet the needs of the particular patient. The line is then connected to the delivery system in use.

Also known as

Oxygen units, O_2 concentrators.

Related devices

Oxygen analyzers, gas regulators.

Where found

Some hospital areas where wall oxygen is not available, such as extended care units; oxygen concentrators are also often used in patient's homes.

Patient Lifts

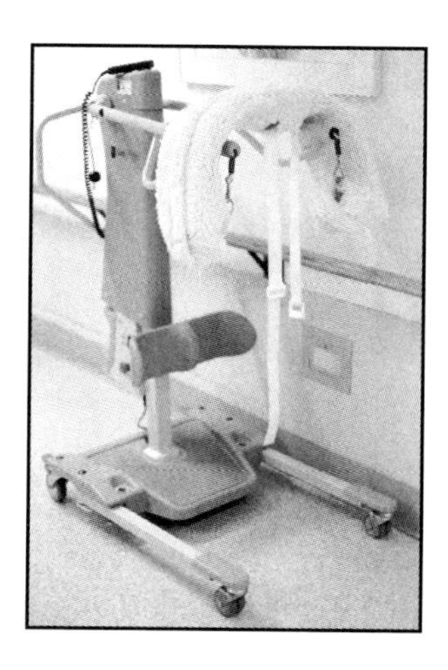

Overview

Patients who are unable to move themselves must be lifted regularly, for treatments, baths, bedding changes, and transfers to wheelchairs, among other reasons. Caregivers who are involved with these movements must be careful to avoid injury to the patients or to themselves, especially if there are not enough people available to safely move the patients manually.

Function

Patient lifts, as their name implies, are designed to help with such patient transfers. Generally, lifts consist of a metal frame with an arm that can be extended over the patient, a sling or platform that can be slipped under the patient which is then attached to the overhead arm, a base that fits under the patient's location to allow lifting without overbalancing, and some mechanism to raise the arm once the patient is secured. There are wheels to allow the lift to move around, and locks on the wheels to keep it steady while the patient is being raised or lowered.

Some lifts have the ability to move the patient horizontally with respect to the base, but this must be limited to within the balance area of the base. Lifting is accomplished by a manually-operated lever/jack mechanism or a battery-powered electric motor. With electric lifts, there must be provision to lower the patient manually should the motor or battery fail.

Lifts must be designed with adequate stability to prevent tipping, so the bases must be quite long and wide. Also, there should be some mechanism to prevent attempts to lift patients who are too heavy for the capacity of the lift, such as an alarm or a mechanical lock-out.

Some lifts incorporate a scale in their design, so that staff can keep track of the patient's weight.

Application

The patient is maneuvered onto the sling or chair portion of the lift. Care must be taken not to twist or pinch the patient. The sling or chair connects to the lift arm, then the mechanism to lift the patient is activated. Once at the appropriate level, the patient is then moved to the desired location and lowered carefully.

Also known as

Lift chairs.

Related devices

Infant scales.

Where found

Most areas of the hospital, especially extended care units and spinal cord injury units.

Patient-controlled Analgesia Pumps

Overview

When patients are experiencing ongoing pain, either from surgery, injury, or disease, intravenous analgesia requirements are variable and subjective. Giving the patient some control over when dosages are administered both allows them better control over their pain and allows nursing staff to attend to other, more critical duties.

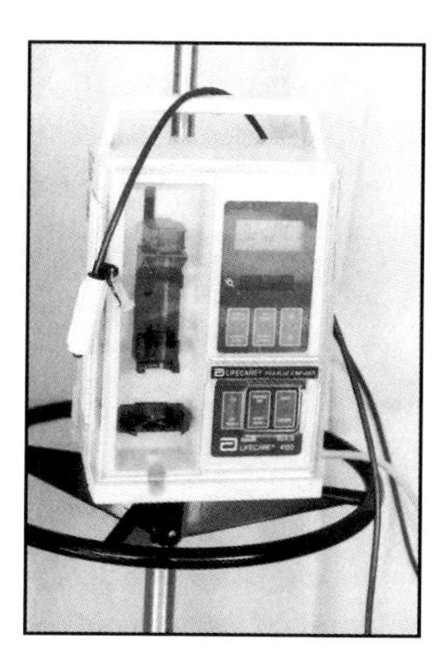

Function

A patient-controlled analgesia pump allows a patient to control his own pain-killer administration. To help prevent abuse of analgesic drugs, the solutions are usually packaged by the hospital pharmacy in sealed glass cartridges. These cartridges contain a piston and delivery port that is compressed by the pump mechanism to deliver the solution. This mechanism is precisely controlled to ensure accurate dose delivery.

Pumps can be programmed for continuous delivery rates, maximum total dosages, maximum patient-controlled dosage and maximum patient-controlled dose frequency. Most units have pre-programmed values for specific drugs at specific concentrations, which can be selected directly, thus avoiding some potential calculation errors. Most controls are located behind a lockable panel.

Application

An intravenous line is established, and staff sets the various parameters so that harmful dosages cannot be delivered. The control panel is then locked, and the patient is given a cable with a push-button on the end, which, when pressed, causes the pump to deliver the pre-set dosage of medication.

Also known as

PCA pumps.

Related devices

IV pumps.

Where found

Post-operative recovery areas, palliative care units, burn units, other patient areas where pain management is necessary.

Physiological Monitors

Overview

Patients in serious condition must have their vital signs monitored continuously.

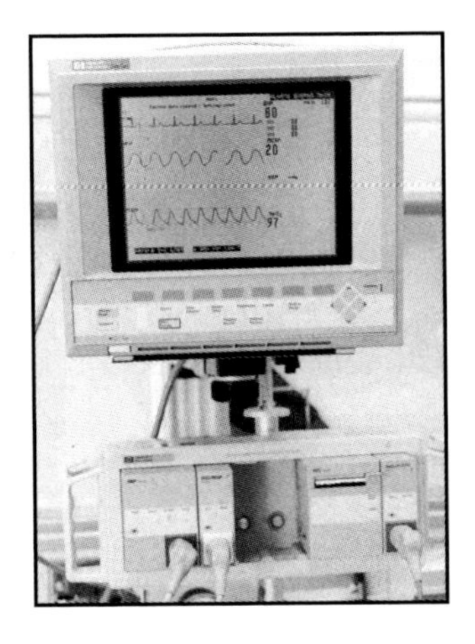

Function

Physiological monitors bring a number of different critical measurements together so that a comprehensive picture of the patient's condition can be obtained. The number of parameters measured is different for various situations, and monitors are designed to meet these different needs.

There are two general formats of physiological monitors: configured, in which the unit is built with various measurement capabilities and cannot be changed; or modular, in which various modules that measure different parameters can be added or removed as needed. Configured monitors are typically more compact and less expensive; modular monitors are more flexible in terms of the types of measurements they can perform, and have the advantage that, if a particular module fails, it can simply be unplugged and exchanged with another in minutes.

Monitors can measure ECG, respiration, temperature, blood oxygen saturation (SpO_2), blood pressure (either non-invasively or invasively), other pressures (such as intra-cranial or spinal), temperature, cardiac output, exhaled carbon dioxide levels (capnography), blood carbon dioxide saturation, and blood chemistry (POC blood analysis). In addition, monitors may be able to interface with other devices such as ventilators and nurse call systems. They may have a recorder module included, or they may connect to a central monitoring system and/or patient charting system.

The display portion of a monitor shows the results of the various measurements, either as a graph or as a numerical value, or both. Monitors can usually display a number of different parameters simultaneously, and can store values even when they aren't displayed.

Most parameters monitored can be set up so that an alarm sounds if the values go beyond certain pre-determined levels; these alarm levels are set to generic values when the monitor is turned on, but can be changed to suit individual circumstances. Alarms may be audible or visual or both, and they may be set to initiate a recording, signal a central monitoring station, or trigger a nurse-call system.

Monitors may record various measurements, which can later be called up as tables or graphs so that changes can be tracked more easily; some systems allow marking of specific events, such as physical activity or medication administration, so that these can be correlated to measurements. Tables and graphs may be available for print-out on a built-in recorder or on a larger-format printer at a central station.

Some monitors have a built-in battery so they can continue to function if there is a power failure or during patient transport.

Application

Monitor application is complex, and varies from one model and/or manufacturer to another. A user's manual and/or experienced operator must be consulted before setting up a monitor.

Generally, the various sensors to be used are applied to the patient, calibrations are performed as necessary, alarm levels and other situation-specific parameters are adjusted, and monitoring commences.

Also known as

ECG monitors, bedside monitors, monitors, multiparameter monitors, patient monitors, cardiac monitors.

Related devices

ECG recorders, SpO_2 monitors, NIBP monitors, invasive pressure monitors, capnographs, defibrillators, central stations, probe thermometers, point of care blood analysis systems.

Where found

Intensive care areas, emergency rooms, operating rooms, special treatment areas.

Point of Care Blood Analysis Systems

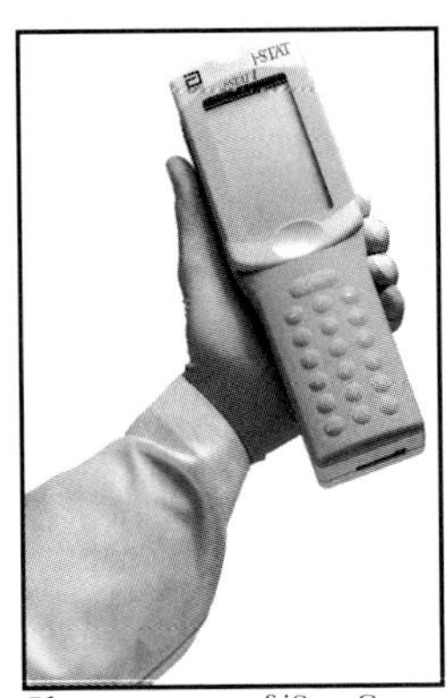

Photo courtesy of iStat Corp.

Overview

In order to provide the most timely and effective treatment of patients, caregivers need to be able to assess their condition as completely as possible. An important aspect of this assessment is the patient's blood chemistry. Levels of blood gases (especially oxygen and carbon dioxide), electrolytes (sodium, calcium, potassium), and other components (urea, hemoglobin) can fluctuate rapidly in response to injury or disease processes or as a result of medications or other therapies, such as oxygen administration or temperature control. In the past, determination of blood chemistry involved calling a technician to take a sample, having the blood drawn, sending to the laboratory, having the analysis performed, obtaining a list of measurements, and returning the list to the patient's bedside.

While in many cases this is an adequate process, critical patients need to have blood chemistry results available more rapidly in order to ensure the most effective treatment.

Function

Point of care analysis systems have been developed to meet the need for quick results. Generally, they take one of two forms: a stand-alone unit with its own display screen and printer; or a module that connects to a physiological monitor, in which case the monitor's display and recorder are used.

The function of these systems depends on specially-designed cartridges (specific to one or sometimes two or three blood components), into which a small blood sample is placed. Chemicals within the cartridge are applied to the sample, and the ensuing reactions produce an electrical signal which can then be measured by the main unit; this signal correlates to the particular blood component being analyzed. The results are displayed on a screen and printed out; results may be tabulated for a printout or graphed so that changing values can be tracked.

A number of different cartridges must be kept on hand, in specific environmental conditions, and the systems must be carefully quality-controlled to ensure proper measurements. The range of measurements is somewhat less than can be obtained from a full laboratory analysis, but blood samples required are much smaller (a few drops that can be taken from a finger prick, as opposed to a syringe full from a blood vessel for lab testing) and results are available much more quickly, usually within a few minutes of the sample being taken. Also, if the blood analysis unit interfaces with the patient monitoring system, measurements can become part of the overall patient vital signs record, allowing a more complete view of their current and past condition.

Some new systems can analyze a more limited range of blood chemicals using a high-tech tip on an in-dwelling catheter, producing even faster measurements without the need to take blood samples at all.

Application

The unit is calibrated, and then a cartridge is selected, depending on the values to be measured. A small blood sample is obtained, usually from a finger-prick, and is applied to the cartridge, which is inserted into the measuring port. After a pre-set time, the measured values are displayed.

Also known as

POC systems, blood analysis units, chemistry analyzers.

Related devices

Physiological monitors, SpO_2 monitors, ventilators, capnographs, blood glucose analyzers.

Where found

Intensive care units, emergency rooms, operating rooms.

Pulse Oximeters

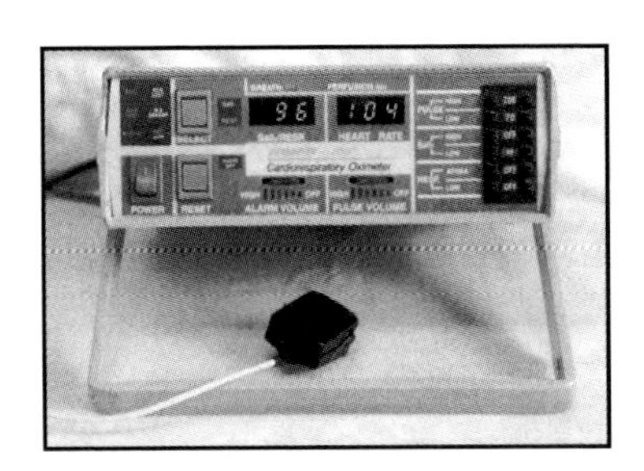

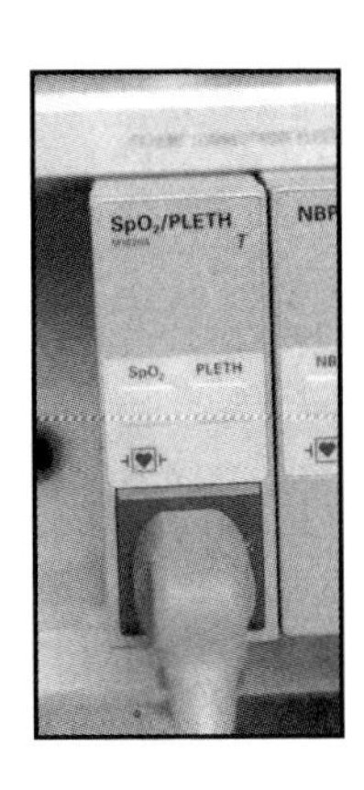

Overview

Many disease conditions can reduce the amount of oxygen present in a patient's blood. Physical effects of low oxygen levels may not been apparent until after the optimal time for intervention, therefore a means of measuring oxygen content (or oxygen saturation) of the blood is important.

Function

A pulse oximeter measures the amount of oxygen in the patient's blood by sending both red and infrared light beams through tissue, and measuring how much of each is transmitted. Blood in the tissue transmits light differently depending on its oxygen saturation levels; by comparing the values for infrared and red light transmission, the unit can calculate oxygen percentage saturation.

Most oximeters also give a read-out of pulse rate, in beats per minute. Some have a tone that sounds with each pulse beat, which may change in pitch depending on the oxygen concentration. Some may also have alarms for high and low oxygen levels and pulse rates; the alarm levels may be fixed or adjustable, though adjustable alarms are usually only used in more critical settings.

Application

A sensor, either wrap-around or clip-on, is placed on a small part of the body that has good blood flow, such as a finger, toe, or earlobe. The sensor has an emitter that produces the red and infrared signals, and a receiver that picks up what is left of each after they pass through the tissue. The unit then processes the measurements and gives a digital readout of the saturation value.

Also known as

O_2 (oxygen) sat meter, sat meter, oximeter, pulse-ox, SpO_2 meter, SaO_2 meter.

Related devices

Oxygen analyzers, ECG monitors, physiological monitors, oxygen concentrators.

Where found

Most areas of the hospital.

Sphygmomanometers

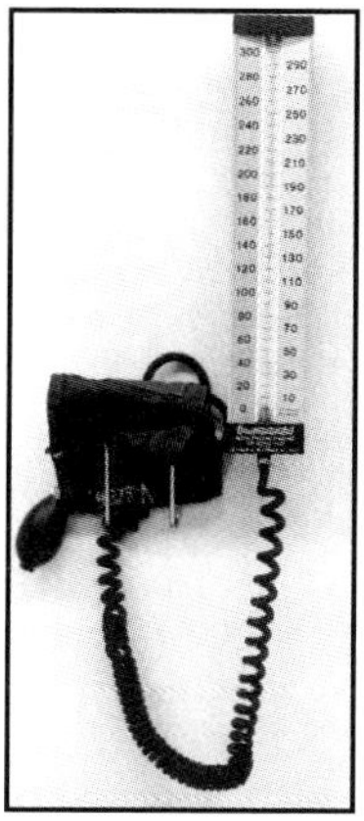

Overview

Blood pressure is a critical vital sign.

Function

An indication of blood pressure can be gained using a sphygmomanometer, which consists of an inflatable cuff, a stethoscope, and a pressure gauge, or manometer. The cuff is placed around the patient's limb (usually the upper arm, but sometimes the thigh or other parts), and is normally held in place with attached Velcro strips. A rubber bulb is used to inflate the cuff while the practitioner listens to pulse sounds in an artery distal to the cuff. When not occluded, there is little noise from the artery. When the cuff pressure reaches diastolic pressure, a sound is produced by the alternate starting and stopping of blood flow. This continues until the cuff reaches systolic pressure, at which point the sound ceases again because there is no blood flow at all. Pressure is indicated by either a graduated mercury column, a mechanical dial, or a simple electronic pressure indicator.

It is usually easier to determine the points where sounds start and stop when the cuff is being deflated by opening a valve, as the drop is smoother than the rise under pumping and there is no interference from the pumping itself.

Application

A cuff of the appropriate size is placed on the patient's limb. The practitioner listens to blood flow sounds distal to the cuff, and the cuff is inflated until sounds stop, indicating complete occlusion of the artery. Pressure is slowly released, and the pressure at the beginning and ending of flow sounds is noted and recorded.

Also known as

BP cuff, BP unit, blood pressure units, manual BP units, pressure cuff.

Related devices

NIBPs, invasive pressure monitors, stethoscopes.

Where found

Most patient care areas of the hospital. Since many areas will have electronic blood pressure units, the manual ones might not be used much there, though they will usually be in place if needed.

Stethoscopes

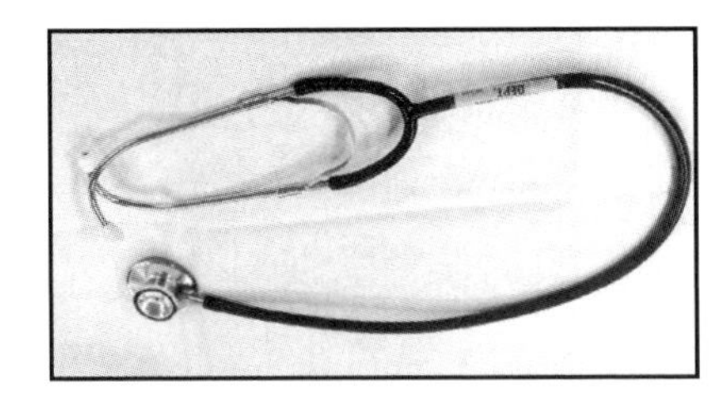

Overview

In many circumstances, sound is important in determining the state of a patient's health. Heart beat, lung condition, blood flow, gastrointestinal function, and fetal health, among others, are all parameters that produce sounds that can aid in diagnosis.

Function

The stethoscope is one of the oldest of medical devices, the earliest being simply a cone that magnified sounds and carried them to the physician's ear. Most modern stethoscopes still operate on this principle, though refined and perfected. A cup- or bell-shaped part picks up sounds from within the body. The end of this part may be either open or covered with a diaphragm, which helps to transmit sounds in some circumstances. Some stethoscopes have two different-sized cups for different situations. The sounds are then carried by hollow tubes, divided by a Y, and carried to each ear. Some units have disposable earpieces, especially if more than one person regularly uses them.

Electronics technology has allowed the stethoscope to become more sensitive and precise, though, of course, an electronic version will be more expensive (and usually less durable) than its mechanical counterpart. Electronic stethoscopes use a microphone to pick up sounds. The microphone may be in a bell-housing, as with mechanical stethoscopes, or the microphone element may be applied directly to the patient. In the latter case, a gel may be used between microphone and skin to help increase sound conduction and reduce outside interference. The signal from the microphone is amplified and filtered, then fed to a small speaker which may be open so that more than one person can listen, or inside a structure much like the Y and earpieces of mechanical stethoscopes and carried to both ears of the practitioner. Electronic amplification also allows filtering of extraneous signals and precise volume control.

Application

The head of the stethoscope is placed on the patient's skin near to the area whose sounds are of interest.

Also known as

Steth.

Related devices

Fetal heart detectors.

Where found

Most areas of the hospital.

Syringe Pumps

Overview

Some intravenous medications must be delivered in relatively small quantities but in very accurate dosages.

Function

In such cases, a syringe pump is often used. These devices consist of a motor and drive system, into which a standard hypodermic syringe is fitted. The motor presses the plunger of the syringe, delivering the medication according to the rate and dosage information programmed by the operator. Syringe pumps are sometimes used in ambulatory situations, such as for pain medication, chemotherapy, or chronic conditions requiring small quantities of medication, such as insulin.

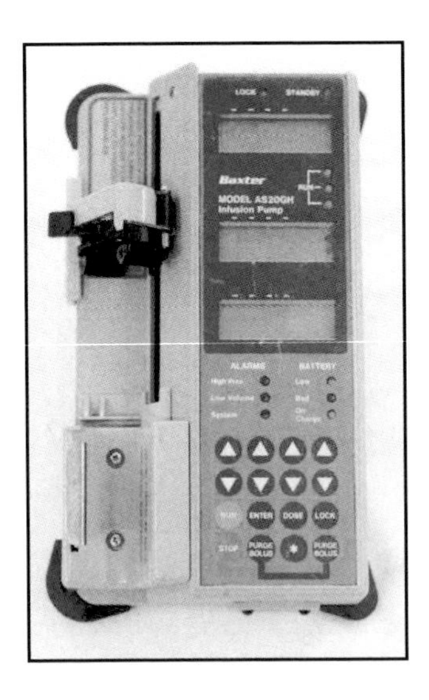

Application

An intravenous line is established, and a syringe is loaded with the medication to be delivered and placed into the cradle of the pump. Flow and dose parameters are set, and the pump is started.

Also known as

(None)

Related devices

IV pumps, PCA pumps.

Where found

Operating rooms, critical care areas, palliative care units.

Tympanic Thermometers

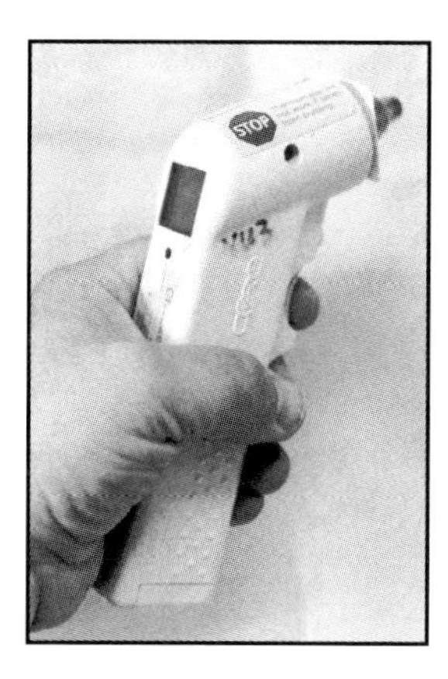

Overview

Medical caregivers need to know the patient's internal temperature, and want to be able to measure it quickly, with minimal discomfort to the patient.

Function

A tympanic thermometer does this by measuring the infrared (heat) radiation of the eardrum (or tympanic membrane, which is where this device gets its name). This is usually very close to internal temperature.

Application

After applying a disposable cover (which goes over the tip to help prevent the passing of material from one patient to another, and also to keep the tip clean), the sensor of the device is inserted into the patient's ear canal; it is important, though sometimes difficult, to have the sensor pointing directly at the eardrum. Otherwise, the temperature of the wall of the ear canal is measured, which may be different than the actual core temperature. When the probe is positioned correctly, a button is pressed to initiate a reading; when the measurement stabilizes (sometimes indicated by a sound), the value is displayed in either Centigrade or Fahrenheit degrees. This technique requires some practice to produce useful results.

It should be noted that these devices compare the temperature of the target with that of an internal heat source. Therefore, the unit must be near room temperature to function properly. Also, the clear window at the end of the sensor must be clean; any obstruction affects accuracy.

Also known as

Infrared thermometer, IR thermometer.

Related devices

Electronic probe thermometers, glass tube thermometers.

Where found

Most areas of the hospital.

Ventilators

Overview

In response to disease conditions or trauma, the ability of a patient to breathe may be reduced or non-existent. In these cases, ventilation must be provided artificially, either through mouth-to-mouth means, via a 'breathing bag' which is pumped by hand, or by a mechanical ventilator. For situations where artificial ventilation is required for more than a few minutes, a mechanical ventilator is preferred.

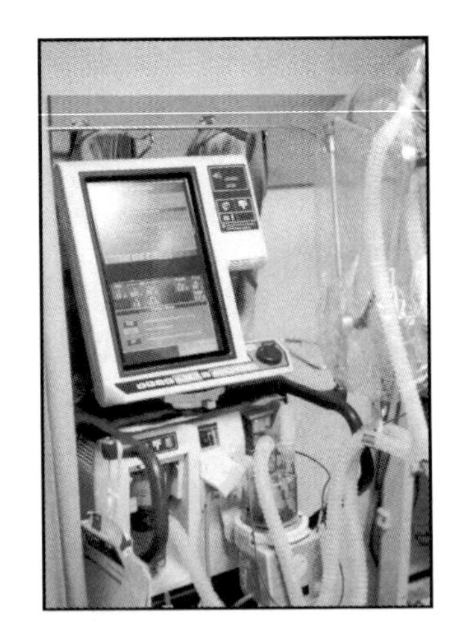

Function

Ventilators range from relatively basic devices that simply provide a properly-timed boost in air pressure to assist patients in drawing air into their lungs, up to very complex units with a number of variable parameters, built-in compressors and oxygen blenders, monitoring and alarms systems, and battery back-up power.

The goal of all systems is the same: to provide adequate ventilation to the patient, while minimizing harm to their lungs and associated structures. Minimizing harm is especially important in situations where ventilation may be required for periods of days, months, or years.

Most full-featured ventilators can operate in a variety of modes, depending on the needs of each patient. Breaths may be delivered by the ventilator at a pre-selected rate; when the patient goes too long without taking a breath unassisted; or whenever the patient makes an effort at drawing a breath (within set limits).

Since ventilation is obviously a critical factor, ventilators must be designed to be highly reliable, in both normal and emergency situations. Ideally, they should be able to continue to function if line power or wall-oxygen pressure should fail, and should have redundant critical components to minimize the risk of failures. Alarms must also be reliable, and must be designed so that they cannot be turned off.

When a patient is being ventilated, it may be important to monitor the oxygen content of delivered air, blood oxygen concentration, and expired CO_2 levels. These functions may be performed by separate devices or by ones integrated into the ventilator; CO_2 levels are usually measured by a separate device. All of these functions may be performed by a physiological monitor; some such monitors can interface with the ventilator, allowing for recording of ventilation parameters and integration of alarms into a central monitoring system.

Application

Ventilator function is very complex and critical, and varies greatly from one model and/or manufacturer to another. An experienced operator must be consulted during setup and use.

Generally, an airway is established, either with an endotracheal tube or via a tracheostomy, and then the ventilator is connected. After setting the various parameters to suit the patient, the unit is turned on, and it begins to breathe for the patient.

Also known as

Vents, respirators, breathing machines.

Related devices

Oxygen analyzers, capnographs, physiological monitors.

Where found

Critical care areas.

2. Cardiology/Respiratory

In some hospitals, cardiology and respiratory are separate departments. In others, they are combined, or they may even be integrated into another area, such as the laboratory or the physiotherapy department.

Devices typically found in these areas, in addition to the equipment listed in this section, include: defibrillators, ECG machines, manual or non-invasive blood pressure units, pulse oximeters, stethoscopes, aspirators, and ventilators.

Ambulatory ECG Recorders

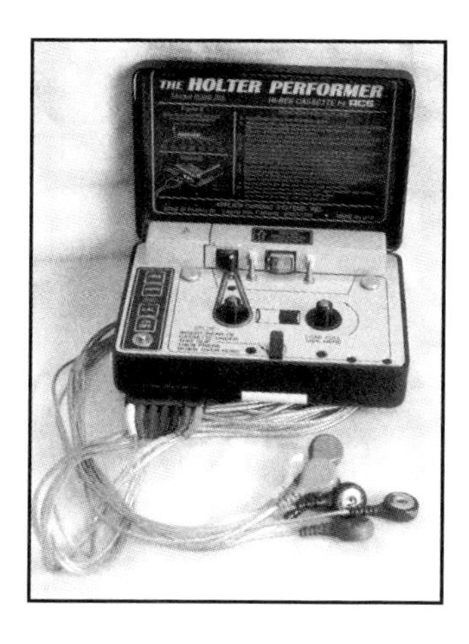

Overview

It can be important to track the ECG signals of patients for an extended period of time, as they go about their daily activities.

Function

Ambulatory ECG recorders perform mobile ECG measurement, by putting the input and processing circuitry for ECG signals into a small, battery powered package, along with a means of recording the signal. In the past, this was done with magnetic tape (often a standard cassette, geared down to run for 24 hours), but newer units use electronic memory modules which have the advantage of needing no moving parts or mechanical alignment, thus, being lighter, more durable, and requiring less maintenance. Picking up the ECG signals from the patient involves the same type of skin electrode and lead wires as other ECG monitors and recorders, but since there is no need for display or print-out of the signal in the device itself, the large and high-power-consumption monitor/recorder can be eliminated. Also, since the signal is not analyzed on the machine and there are no alarms or other parameters to be considered, the electronic circuitry can be relatively simple.

These units have a clock that records a coordinating signal along with the ECG, and there is normally a button that the patient can use to mark times of exertion, chest pains, meal times, or other circumstances as requested by their physician.

Application

Electrodes are placed on the patient's skin in prescribed locations, and the device is connected and started. To ensure the proper functioning of the device at the start of recording, there is usually a port that connects to the circuitry after amplification and processing, allowing the operator to check that everything is working at least to that point. The unit is worn in a comfortable pouch that allows access to the 'mark' button while protecting it from the elements. The patient is asked to keep a diary of activities and symptoms, to complement the 'marks' made by pressing the button. Since timing is an important factor in these recordings, the units have an internal clock which must be checked and set, if necessary, before recording starts.

After the recording period (usually about 24 hours), the signal is read by an analysis machine that can display the whole track on a video screen or print it out on a recorder. Clinically-significant sections of the trace can be magnified and viewed or printed in more detail, and the analyzer can perform various measurements and analyses, such as average, minimum, and maximum heart rates, frequency of abnormal beats, and others. The signals can be correlated to the diary and the mark signals to help in understanding the condition of the patient's cardiac system.

Also known as

AECG machines, long-term ECG recorders, Holter recorders (after one of the first manufacturers of this type of device).

Related devices

ECG monitors, ECG recorders, cardiac stress test systems.

Where found

Cardiology Department, Laboratory Outpatient Department.

Cardiac Catheterization Systems

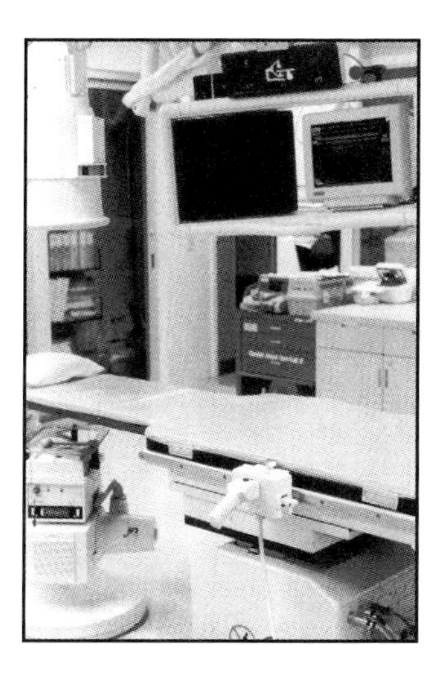

Overview

The diagnosis of certain heart conditions can be aided by visualizing the coronary arteries.

Function

Cardiac catheterization ('cath') systems allow this visualization by injecting contrast medium (radiopaque dye) into the target arteries, then taking x-rays as the dye moves through the vessels.

The system is made up of several components. A monitoring system measures, displays, and records pressure values from the catheter, along with the patient's ECG signal. X-ray apparatus is required to visualize the progression of the catheter and the movement of the injected dye (recorded on video tape or other storage medium for later analysis). There is specific equipment for handling the catheter itself. Finally, a dye injector delivers the contrast medium. Since it is a critical procedure, a defibrillator is normally kept on hand as well.

The injector must be capable of delivering precise quantities of dye very quickly, which means that high pressures are involved.

Application

Operators insert a catheter into the heart (usually from a point in the patient's groin), through the aortic valve and into the coronary artery, and then inject a radiopaque dye; the catheter is then withdrawn. X-ray images are taken and ECG and pressure measurements are made throughout the procedure.

Also known as

Cath lab.

Related devices

ECG monitors, invasive pressure monitors, defibrillators, dye injectors.

Equipment location

Specialized cardiac catheterization lab, which may be in a cardiology department, in radiology, or in a separate unit.

Cardiac Output Systems

Overview

One of the ways of determining a patient's state of cardiac health is to measure the amount of blood pumped by each cycle of the heart. This value is the cardiac output, and though it cannot be measured directly, techniques have been developed to determine this value very accurately.

Function

There are two basic methods of measuring cardiac output. One uses colored dye; the other uses cold saline solution. Both rely on the same principle.

In each case, a double-walled catheter is inserted into the patient's venous system and guided to the vena cava, through the heart, and into the pulmonary artery. An injectate opening is located a precise distance from the end of the catheter, in the right atrium of the heart; on the end of the catheter, in the pulmonary artery, is a sensor.

In the dye dilution method, a specific amount of dye is injected through the catheter into the right atrium at a specific rate. As blood flows through the vessel, it mixes with the dye, diluting it. When the diluted mixture passes the tip of the catheter, a special optical sensor measures the concentration of dye. This measurement is done for a period of time, and the results are entered into a computer. A graph is obtained of dye concentration versus time, and by performing calculations on the measurements, an accurate measure of blood flow over time can be determined; this value is the cardiac output.

The cold saline method is similar, except the saline substitutes for the dye, and a temperature sensor is used in place of the optical sensor. Saline is less likely to produce side effects in the patient than dye, and the temperature sensors tend to be more stable than the optical probes.

Either method can be performed by a stand-alone device, or by a module that is part of a physiological monitoring system. Stand-alone units need a display for prompting injection times, etc., and showing results; they also usually have a built-in recorder. Modules can use the display and recorder of the monitor system for these functions.

Application

The special catheter is inserted into the patient's venous system, often via the subclavian vein in the neck. It is then advanced until it reaches the correct location, determined by observing pressure measurements taken from the catheter tip. At a precise time, the injection is started. Temperature or dye concentration measurements are taken throughout the procedure. The process may be repeated, though too much dye or cold saline can cause difficulties. When satisfactory results have been obtained, the catheter is withdrawn.

Because most cardiac output measurements are performed on patients with heart problems, and because the procedure can cause side effects such as cardiac arrhythmias, staff must be well-trained and alert, and emergency equipment such as defibrillators must be immediately available.

Also known as

(None).

Related devices

Physiological monitors, electronic probe thermometers.

Equipment location

Intensive/cardiac care units, cardiology labs.

Spirometers

Overview

Measuring lung function is critical in determining the state of health of many patients, especially those with respiratory disorders. Various parameters can be measured, including those of normal breathing and the extremes of inspiration and expiration.

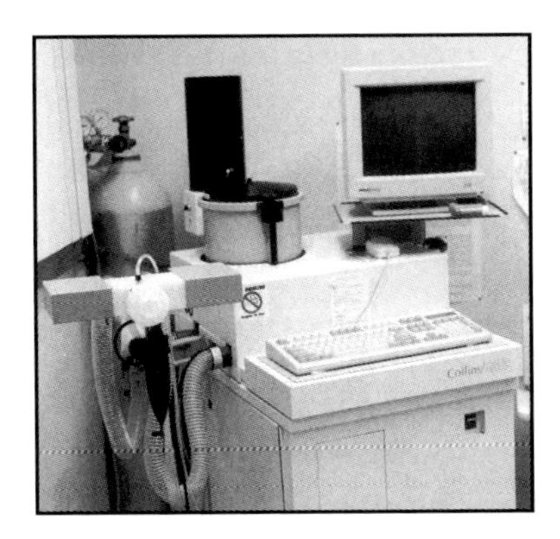

Function

Both the volume and the rate of airflow during breathing can be measured by a spirometer; both can be graphed against each other to produce a roughly circular chart, beginning and ending at the same point, or volume can be graphed against time to give a linear chart. Measurements are usually repeated a few times for each level of breathing (normal and maximal effort). The shape of the charts, including minimums and maximums, as well as any irregularities, can tell the caregiver a great deal about the health of the patient's lungs. In addition, tests performed both before and after medications can give a good indication as to the effectiveness of the medication. Long-term tracking of test results provides a means of following disease and/or recovery progress.

The simplest spirometers are strictly mechanical, with the patient breathing into a bellows that has a chart pen attached, which in turn marks rotating or moving graph paper, producing a linear chart of volume versus time. More advanced versions use some kind of electronic sensor to measure flow; once the flow rate and cross-sectional area of the breathing tube are known, volumes can be calculated. These data can then be used to produce the circular graphs mentioned previously, either on a graph paper or on a video monitor. Electronic circuitry can perform all relevant measurements and give a print-out of the results, as well as allowing pre- and post-medication tests to be compared directly, on a graph and/or as numerical values. Spirometers may also form part of a system that can monitor such parameters as expired CO_2 levels, body temperature, ECG, blood oxygen saturation (SpO_2), and others.

The prime considerations of spirometer design are that they be accurate and repeatable, and that they provide an absolute minimum of obstruction to the patient's breathing.

Application

A fresh mouthpiece is inserted into the breathing tube of the unit, and the patient places it in the mouth. Normal breathing is done for several cycles, and then the operator asks the patient to take in as much air as possible, hold it for a second, and then exhale that breath as thoroughly as possible, as rapidly as possible. The operator examines the results, and may ask the patient to repeat the procedure, sometimes giving specific instructions to help obtain a better result.

If pre- and post-medication studies are being done, the patient is given medication and the test repeated, sometimes at a few pre-determined times after medication administration.

Avoiding disease transmission is, of course, an important consideration, and all spirometers have disposable mouthpieces, which are changed after each patient. In addition, parts exposed to patients are sterilized at regular intervals.

Also known as

Spirometers, pulmonary function testers/machines.

Related devices

Stress test systems, pulse oximeters, capnographs.

Equipment location

Respiratory clinics, some doctors' offices, laboratories (usually in smaller hospitals), mobile respiratory testing facilities.

Stress Test Systems

Overview

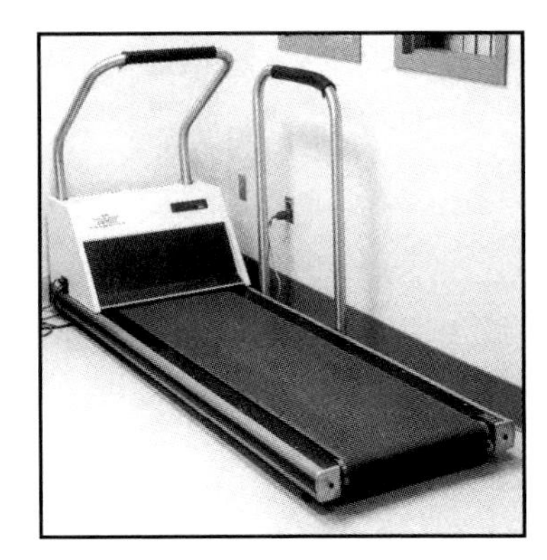

The ECG of a patient is usually measured with the patient in bed. In diagnosing many cardiac problems, however, the physician needs to see how the heart reacts to exercise.

Function

Stress test systems allow the opportunity of testing ECG during exercise. They consist of a mechanism for exercise in a fixed location, such as a treadmill, and a component that combines control circuitry for the exercise device with physiological monitoring. The monitor looks at the patient's ECG waveforms, usually on multiple leads, and displays the signals on a screen and on a chart recorder. Some systems may also monitor other parameters, such as oxygen saturation (SpO_2, pulse oximetry) or respiration.

The system controls the level of exercise according to a pre-programmed pattern. For example, by varying speed and/or slope of the belt of a treadmill, exercise levels can be changed. The exercise levels can also be controlled directly by the operator; however, the preset programs are usually used. These are developed by cardiologists who have determined the optimum exercise patterns for various patients and conditions, in order to maximize the chance of obtaining clinically useful results while minimizing risk to the patient. The operator can stop the exercise program at any time if it appears that the patient is becoming over-stressed, and systems have safety mechanisms to stop if the patient moves off the equipment during exercise.

ECG signals are picked up from the patient's skin just as with normal ECG monitors, but with certain modifications. Because the electrodes will be applied for a short time, but must be able to withstand (sometimes vigorous) patient movement and possible perspiration, they are designed somewhat differently than those used for longer-term monitoring. Suction cup electrodes may be used, with rubber bulbs to create suction and silver cups to make contact with the patient's skin. A conductive cream is usually used with this type of electrode. A harness of some sort may be used to keep the electrode lead wires and cables in place, and, of course, the cable must be quite long to allow adequate patient movement.

Application

After skin preparation, electrodes are applied to the patient's skin. They may be held more securely in place with a harness or vest. The patient moves onto the exercise machine, and the operator initiates the procedure. While the operator (and sometimes the monitoring system) watches for signs of problems, the control system runs through the pre-set exercise program, which usually includes a slow 'warm up' phase, a long 'high-intensity' phase, and a 'cool down' phase.

Recordings may be made continuously during the exercise program, or for shorter periods, either at pre-selected times or when the operator decides something should be examined more closely.

Since many of the patients undergoing this type of testing have cardiac problems, a defibrillator is normally kept nearby during procedures.

Also known as

Exercise systems, treadmills.

Related devices

ECG monitors, ECG recorders, pulse oximeters, spirometers, defibrillators.

Equipment location

Cardiology, outpatient, physiotherapy, or respiratory therapy departments; laboratories.

3. Emergency Room

Emergency rooms must be equipped to handle a wide range of medical situations, from extreme traumas and critical disease conditions to fevers and splinters. The range of equipment found in this area is therefore greater than in any other part of the hospital.

In some hospitals, the emergency room may be integrated with or closely connected to the outpatient department. Devices typically found in emergency rooms, in addition to the equipment listed in this section, include: aspirators, capnographs, ECG machines, electronic probe thermometers, tympanic thermometers, examination lamps, gas regulators, glucometers, intravenous pumps, invasive pressure monitors, non-invasive pressure monitors, oto/laryngo/ophthalmoscopes, physiological monitors, point of care blood analysis systems, sphygmomanometers, stethoscopes, ventilators, defibrillators, cardiac pacemakers, oxygen analyzers, cast cutters, slit lamps, and C-arm units.

Blood Warmers

Overview

Blood must be kept refrigerated to prolong its storage life, but infusing it into a patient at this temperature can cause a serious drop in body temperature, especially if the patient is small and/or hypothermic, or if a large amount of blood is required in a short time.

Function

A blood warmer is a device that allows blood to be heated to near body temperature before it is infused. This requires careful control, however, as overheating blood can damage it. Blood warming units use either a water bath or metal plate heaters to warm the blood. Since the temperature must be increased substantially in a short time, there must be a large surface area for sufficient heat exchange to take place. In metal plate heaters, the blood passes through a plastic pouch that has a long, back-and-forth passage through it, so the blood has a long distance in which to be warmed. Water bath units have the advantage of quicker heat exchange, often using a double-walled tube, with blood flowing toward the patient in the inside tube and warm water being pumped in the opposite direction in the outer jacket. The opposing flow means that the blood is close to the water temperature when it exits the double-walled section; the total length of tubing within the warmer is much less in this type, which means less wasted blood.

In both types of warmers, the exit of the warmer must be as close to the patient as possible so that it doesn't cool off too much before reaching the patient. With its somewhat moveable section of double-walled tube, the water jacket warmer allows for a shorter unheated section of tubing before it reaches the patient.

Both types also usually have a temperature display, as well as a double over-temperature cutout and alarm system, since overheating the blood can be harmful both to the blood and to the patient. The section of tubing in the warmer is discarded after use.

These devices may be used in conjunction with an IV pump, which would be placed upstream of the warmer so that heat isn't lost in the pump and associated tubing.

Application

An intravenous line is established, and the fluid to be administered is connected to the fluid warmer apparatus. Fluid administration is begun, and the liquid is warmed as it passes through the device. The short length of tubing between the machine and the patient ensures minimal temperature drop.

Also known as

Fluid warmers.

Related devices

IV pumps, hyper/hypothermia units.

Where found

Emergency rooms, operating rooms, some special care units.

Hyper/Hypothermia Units

Overview

When a patient's body temperature is significantly higher (hyperthermia) or lower (hypothermia) than normal, it must be returned to the normal range as quickly as possible.

Function

Water immersion provides the fastest heat exchange, either up or down, but this is not always possible or practical. A plastic pad or jacket provides a more flexible and accessible means of delivering heat to or removing heat from the patient.

Some devices perform only one function, either heating or cooling, though some have both functions built into one unit. Warmers, especially air types, are typically much smaller and lighter than cooling units, which require compressors and refrigeration coils. Air has the advantage of quicker heating or cooling, as compared to water, but it also doesn't carry as much heat to or from the patient as water does, though it can flow more rapidly, partially overcoming this disadvantage.

These units normally have a temperature control, as well as over and/or under temperature alarms and cut-offs. They may also have a means of displaying patient temperature. The sensor for this must be placed away from the heat exchange pad in order to give an accurate indication of body temperature.

Application

The heating/cooling pad is placed in direct contact with the patient's skin, covering as much surface area as possible. Temperature-controlled water or air is then passed rapidly through the pad to warm or cool the patient.

Patient temperature and condition must be monitored closely while the unit is being used.

Also known as

Patient heaters/coolers.

Related devices

Blood warmers.

Where found

Emergency rooms, operating rooms, some special care units.

4. Intensive Care Unit (ICU)

For the purposes of this book, intensive care units include areas designated as ICUs, cardiac care units, special care units, critical care units, etc. Because of the intensive nature of medical care in these areas, a wide variety of equipment is found in them.

Devices typically found in these areas, in addition to the equipment listed in this section, include: aspirators, capnographs, cardiac output systems, ECG machines, electric hospital beds, electronic probe thermometers, tympanic thermometers, examination lamps, gas regulators, glucometers, intravenous pumps, invasive pressure monitors, non-invasive pressure monitors, oto/laryngo/ophthalmoscopes, physiological monitors, point of care blood analysis systems, sphygmomanometers, stethoscopes, syringe pumps, ventilators, defibrillators, oxygen analyzers, and C-arm units.

Cardiac Pacemakers

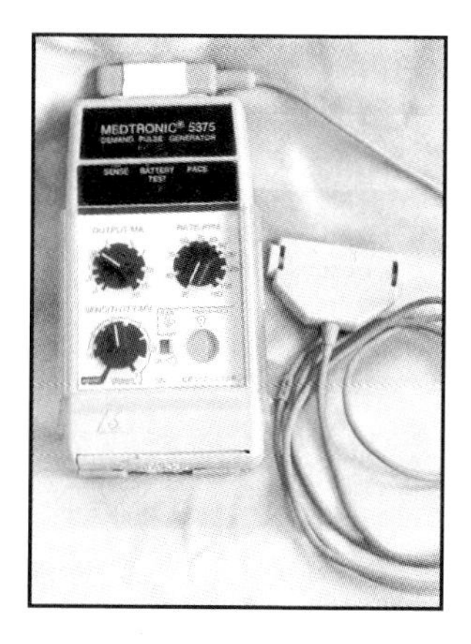

Overview

Cardiac rhythm is normally controlled by a system within the heart, moderated by various parameters such as oxygen demand and levels of hormones such as adrenalin. The natural pacemaker system sends signals to the heart muscle in a pattern that produces coordinated contractions of the various parts of the heart, of a strength and rate appropriate to body state.

Various disease processes can disrupt this natural pacemaker to such a degree that cardiac contractions are no longer sufficient for the needs of the patient.

Function

In these circumstances, an artificial pacemaker is used to provide proper pacing signals. Pacemakers may be temporary or permanent, and may be external or implanted. External pacemakers can be further divided into invasive and non-invasive types.

External pacemakers are typically used for short-term applications, either until the patient's natural pacemaker can resume normal function or until an implantable pacemaker can be installed, with non-invasive types being used for shorter times than invasive.

Non-invasive external pacemakers use electrodes placed in specific places on the patient's chest to pass electrical signals into the heart. These signals stimulate the heart to beat more effectively, and are usually coordinated with whatever natural cardiac signals are present. They can be adjusted for rate and amplitude, and may either completely control cardiac contractions or act as a 'booster', filling in for missing beats as required. The control signals have to be quite large for enough of the signal to reach the heart, and the signal passes through areas of the body where it isn't needed. Long-term use of electrodes on the patient's skin can cause irritation or burns.

Invasive external pacemakers function in a similar way to non-invasive types, except their signals are carried to the heart by wires inserted into the patient's body and attached directly to the heart. They have the advantage of more precise control and require much less power to effect pacing compared to non-invasive types, but they take much longer to apply, as well as carrying the problems associated with any invasive procedure.

Implantable pacemakers perform the same function as their external counterparts, but have many special restrictions. They are inside the patient's body and therefore relatively inaccessible. This means that they must have a long-lasting power source; even though the power requirements are small, they may be needed for years. Special battery technology has been developed, and nuclear power sources have been used. The units must have some means of being controlled without physical contact. In earlier devices, and many current ones, rate and signal strength could be adjusted using a powerful magnet placed near the implant site. Newer units may have mechanisms to measure body needs and adjust themselves automatically according to these needs. Implanting a foreign object within the body has its own set of considerations apart from the pacemaking functions, and these factors are an important part of the design criteria as well.

Application

Contact is made with the patient's heart, either though external or implanted electrodes. Pacing parameters are set by the operator and the unit is started, which thereafter controls the patient's cardiac rhythm.

Also known as

Pacers.

Related devices

Physiological monitors, ECG recorders.

Where found

Cardiac intensive care units, operating rooms, emergency rooms.

Central Stations

Overview

In any critical care area, medical staff needs to be able to monitor each patient, but not necessarily from the bedside at all times.

Function

Physiological monitors and telemetry units are associated with each patient, and send data to a central location where important signals can be displayed. Most central stations display detailed information about a particular patient as necessary, while normally displaying one waveform and numeric values (usually ECG) for all patients simultaneously.

Patient information can be entered (usually with a standard computer-type keyboard) via the central station at admission, and removed on discharge. Staff may be able to adjust alarm levels from the central station, but usually cannot cancel alarms without going to the bedside.

Recordings of information from the various bedsides can usually be made at the central station, either on a strip-chart recorder or on a larger-format recorder or printer (usually used for tables and graphs or more comprehensive reports). Central monitors can be integrated into or connected to patient data recording and analysis systems. Central monitoring also can give the capability of displaying information from one beside monitor while working at another.

Application

Central stations are used as part of a physiological monitoring system.

Also known as

Central monitors, centrals, nursing centrals.

Related devices

Physiological monitors, telemetry systems.

Where found

Intensive and/or critical care areas.

Telemetry Systems

Overview

Cardiac patients are often able to get out of bed and walk around; this may be part of their recovery process, or it may simply be a convenience. In either case, it is important for medical staff to be able to continue monitoring the patient's ECG signals.

Function

Telemetry systems allow patients to be mobile without unwieldy and potentially dangerous long cables. ECG signals are picked up from the patient's skin by electrodes and wires, just as with regular ECG monitoring. The signals are then amplified and processed, again by similar circuitry as that in a bedside monitor. Within the small module carried by the patient, however, is a radio transmitter that broadcasts signals carrying the ECG information. These signals are picked up by an antennae system and processed to extract the original ECG waveform, which can then be displayed on a central monitor. This monitor can usually display signals from several patients simultaneously, and may be stand-alone or part of central monitoring system that handles information from bedside monitors as well. Recording, trending, patient admission information, and alarms are all handled by the central monitor.

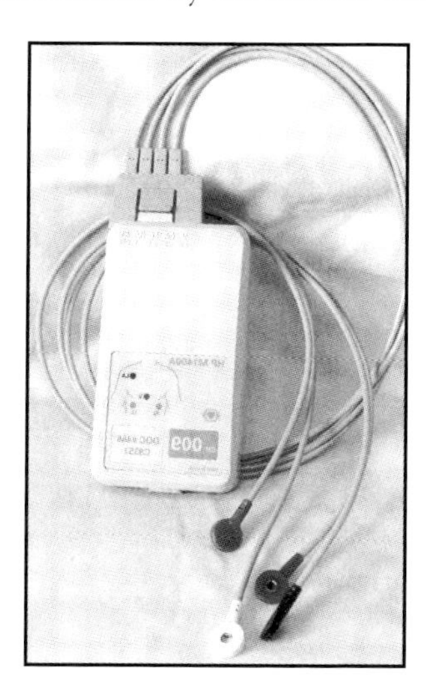

Older telemetries broadcast the ECG information as an analog signal, much like an AM or FM radio station. Newer systems turn the analog signal into a stream of digital information before broadcasting it, which gives better resistance to interference and allows either lower transmitter power (thus prolonging battery life) or else greater range.

Most telemetry transmitters have a nurse call button that the patient can use to signal for assistance or to mark any unusual feelings or symptoms they might have associated with their condition.

Some systems also have the capability of obtaining and transmitting blood oxygen saturation (SpO_2) information as well as ECG, which can help give a more complete picture of patients' condition.

Application

A fresh battery is installed in the transmitter unit, and electrodes are placed on the patient's skin. Electrode wires are then attached to the electrodes and to the transmitter, and the unit is placed in a carrying pouch. Signals are transmitted from the patient to the receiver.

Also known as

Telem, remote monitoring, tele or telly.

Related devices

Physiological monitors, ECG systems, central monitors, SpO_2.

Where found

Cardiac care units, cardiac rehabilitation wards.

5. Maternity

The maternity and nursery areas of a hospital are usually closely associated, though sometimes a special care nursery may be located in a different area. Many maternity units include an operating room for performing Caesarean sections.

Devices typically found in maternity wards, in addition to the equipment listed in this section, include: anaesthetic machines, aspirators, electronic probe thermometers, tympanic thermometers, examination lamps, gas regulators, glucometers, intravenous pumps, non-invasive pressure monitors, physiological monitors, patient-controlled analgesia pumps, point of care blood analysis systems, sphygmomanometers, stethoscopes, defibrillators, and oxygen analyzers.

Apgar Timers

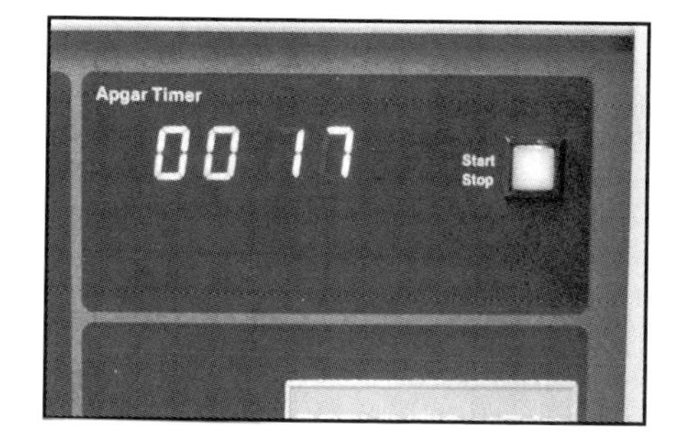

Overview

The condition of newborns can be somewhat difficult to assess. To help in this, Dr. Virginia Apgar developed a set of tests to be performed at specific intervals postpartum, the results of which give a quantitative evaluation of health.

Function

An Apgar timer is simply a device that prominently indicates the times for doing the tests, either audibly or visibly or both. It may be stand-alone or integrated into an infant resuscitation unit.

The observations are made at one, five, and ten minutes after birth, and the tests can be listed using the name Apgar as a mnemonic: appearance (color); pulse (heartbeat); grimace (reflex); activity (muscle tone); and respiration (breathing). Each parameter is rated as 0, 1 or 2 (with 2 being highest) and the scores are added to give a total score at each time. A score of seven or more (out of a possible ten) indicates that the baby is in good condition.

Application

The timer is started when the infant is under observation. The specified parameters are observed and recorded at the times indicated, and the scores are totalled for each time.

Also known as

(None).

Related devices

Infant resuscitators.

Where found

Labor and delivery rooms.

Bilirubin Therapy Systems

Overview

Neonates sometimes have an excess of bilirubin in their system, which causes a jaundiced appearance. It is desirable to reduce these levels, and the simplest method of doing so is to expose the skin to ultraviolet (UV) light, which causes the bilirubin to break down.

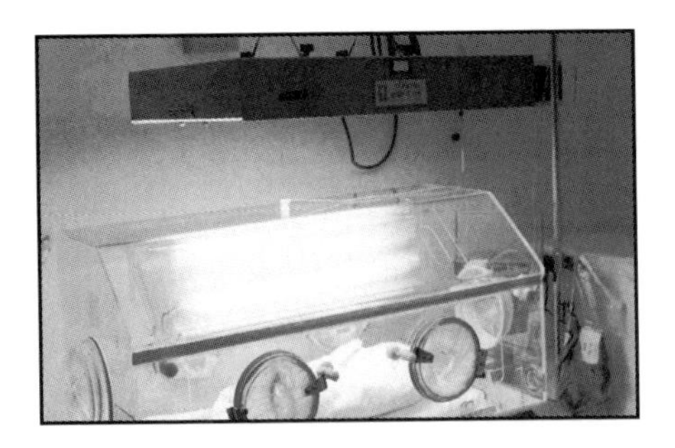

Function

Bilirubin therapy systems are simply light sources that produce ultraviolet light of the optimum wavelength and intensity. Too much exposure can cause burns, while too little is ineffective. Some systems use overhead fluorescent lamps that are designed to emit the correct type and amount of UV light. Several factors affect the amount of exposure the baby receives, such as the skin surface available, material such as the Plexiglass of an incubator between the light source and the infant, the distance between the light and the baby, and the age and condition of the bulbs. Because of these variables, it is important that the UV levels at the baby's skin be measured at the start of treatment, and at intervals if the treatment time is prolonged. In order to prevent excessive exposure, it is also desirable that a timer mechanism be used to either cut off the UV light after a predetermined time or to remind staff to turn the light off. Overhead systems often have a tape measure built in to aid in placing the light at the correct distance from the baby. Since the lights produce some heat, and also because much of the baby's skin is exposed to facilitate treatment, it is important to monitor the baby's body temperature during treatment.

Another type of bilirubin therapy system avoids some of the problems associated with overhead lamp types. This method uses a 'blanket' of material in which fiber optic strands are embedded. These strands carry UV light from a central source and distribute it evenly, so that the baby's skin is exposed to a consistent illumination wherever it is covered by the blanket. The factors of distance and of intervening materials are eliminated, and the light sources used tend to be more stable than fluorescent bulbs. The blanket system is also smaller and less cumbersome than an overheard lamp system. Of course, output levels must be checked regularly and exposure times limited.

Application

After determining the desired exposure, the lamps or blanket are put into the correct position, treatment time is selected, and the light turned on. The infant must be observed regularly during treatment.

Also known as

Bililights, UV lights.

Related devices

Infant incubators, thermometers.

Where found

Nurseries, pediatric wards.

Birthing Beds

Overview

During the process of labor and birth, it is important that the mother is as comfortable as possible, and that she be able to get into a good position for delivery. It is also important that the attending health care personnel are able to access the mother and baby both before and during delivery. Birthing beds are designed to meet these needs.

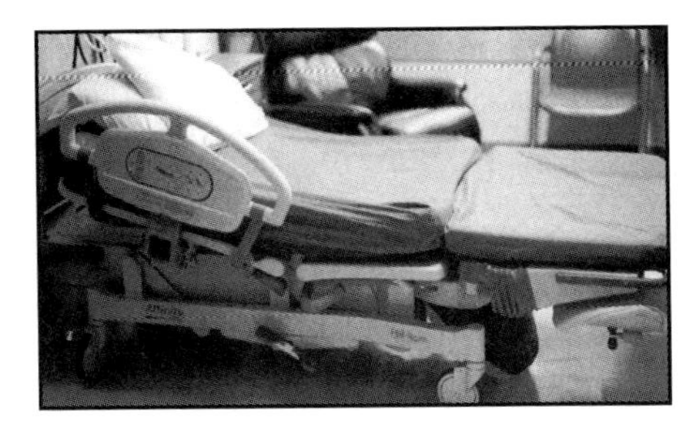

Function

Birthing beds consist of a comfortable, flexible mattress, and a frame and mechanism that allow for a variety of maternal positions. Controls activate motors that raise or lower the whole bed, the head, the feet, and the hips independently (within ergonomic limits); these controls are usually placed both on a pendant or stalk so that the patient or a bedside attendant can use them, and at the foot of the bed for attendant control. Sometimes duplicate controls are on the bed rails on either side of the mattress.

Beds usually have a removable section at the foot to allow caregivers to get close to the birth canal in order to assist in delivery. They generally have channels for optional placement of stirrups, if they will be useful in birthing. As with most hospital beds, the whole unit is on wheels, which can be locked to prevent movement if necessary. Rails on either side of the bed can be raised or lowered, often in multiple positions, to help prevent the patient from falling out of bed. The mattress is covered with an impermeable material to aid in cleaning.

Application

Patients are made comfortable in the bed and shown how to operate the controls. Adjustments are made by the patient, a companion, or a caregiver to suit the needs of the patient. When delivery time approaches, the section at the foot of the bed may be removed to allow attendants access to assist in delivery.

Also known as

(None).

Related devices

Electric hospital beds.

Where found

Maternity ward.

Breast Pumps

Overview

In certain circumstances, infants may need mother's milk at times when it isn't possible for the mother to nurse, such as when the infant is in an intensive care incubator.

Function

Breast pumps provide a means of extracting mother's milk so that it can be stored for later use. They range from simple devices consisting of a suction bulb attached to a collecting cup and reservoir, to somewhat more complex units with electrical pumps. The electric-powered versions usually have controls to vary the rate and/or intensity of suction. A rhythmic action, with periods of suction alternating with no suction seems to work best, as this simulates a baby nursing. The collecting cup and reservoir are similar in all systems, with a cup designed to give a comfortable but snug fit to the mother's breast, and a reservoir that allows the user to see exactly how much milk has been collected.

Application

The collecting cup is applied to the mother's breast, positioned so that the drainage channel is over the nipple, and the pumping action is initiated. With adjustable units, the rate and intensity are modified to give a balance between comfort and optimal milk extraction. Milk which is not to be used immediately is refrigerated.

Also known as

(None).

Related devices

Infant incubators.

Where found

Infant nurseries.

Dopplers

Overview

An important part of prenatal care involves determining the strength and rate of the fetus' heartbeat. This can be done with a stethoscope, but the sounds are often faint, especially earlier in pregnancy, and can be obscured by other sounds from within the mother's abdomen. Doppler probes are more effective for this purpose.

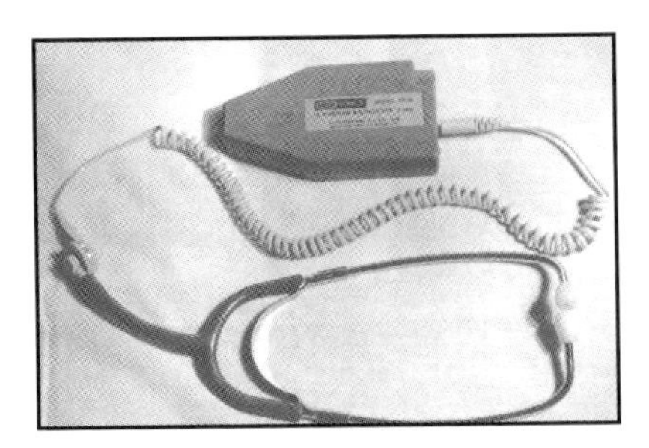

Function

These probes utilize the Doppler Principle, which is that sound waves from a moving source are changed in frequency when the source is moving toward or away from the observer. In this case, sound waves are produced by an ultrasonic probe placed on the mother's abdomen and reflected from the baby's heart. When part of the heart tissue is moving away from the probe, the reflected wave frequency decreases; when moving toward the probe, the frequency increases. The reflected signals are picked up by the probe, and processed in such a way as to produce a sound. The sound is played on a speaker or headphones, and the heart rate and strength can be determined. Some units include a digital readout of heart rate.

Application

A probe is selected, and ultrasound gel is applied to the surface, which is then placed in contact with the mother's abdomen. Power is applied and adjusted as necessary, and the operator scans the beam to locate the fetal heart. When the appropriate signal is heard, position is optimized and the fetal heart-beat is evaluated.

This technique can be very sensitive, and can pick up a fetal heartbeat very early in pregnancy. It requires some skill and patience, however, as the target, especially early in term, can be difficult to locate.

It is possible to focus the probe on the mother's aortic artery, in which case a similar sound is produced; though this usually gives a much lower rate than the fetal heart, care must be take to differentiate the two.

Also known as

Doptones, fetal heart detectors.

Related devices

Fetal monitors, stethoscopes.

Where found

Maternity wards, maternity outpatient clinics, home visit medical kits.

Fetal Monitors

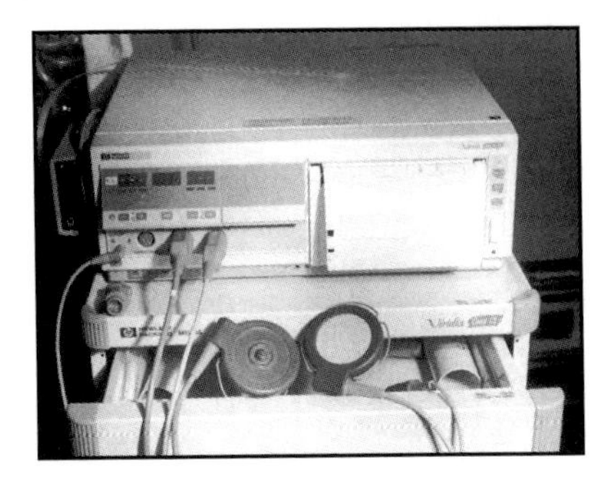

Overview

The time of labor and delivery is critical for the baby, and it can be very useful to be able to monitor its condition, particularly its heart rate. This can be done on a short-term basis with a stethoscope or Doppler unit, but for longer-term monitoring, a fetal monitor is preferred. Fetal monitors can also measure and record other parameters.

Function

Fetal heart rate is picked up in one of two ways. Most commonly, a special transducer is placed on the mother's abdomen. This transducer has several crystals, which produce a beam of ultrasound. The beam is focused at approximately the depth of the fetal heart. When it hits the beating heart, some of the signal is bounced back to the transducer, which picks up this reflection. The frequency of the reflected signal varies as the fetal heart moves in its beating, and this change in frequency can be processed and analyzed to produce a signal corresponding to the fetal heart beat. The rate is displayed as a numerical value, and the sound of the beat is also available at a speaker so that the caregiver can hear if the rate is changing, or if the beam is no longer focused on the baby's heart. The signal is also recorded on a graph chart, which allows caregivers to see trends in the rate over time. In order to allow maximum transmission of the ultrasonic signal, a gel is placed between the transducer and the mother's skin. The transducer is usually held in place by an adjustable elastic belt.

The ultrasound transducer is quick and easy to apply and can give very valuable information. However, it can lose the signal through movements of either the fetus or the mother.

The second method of picking up the fetal heart rate is through an electrode. A curved section of fine silver wire is inserted through the vagina and pierces the surface of the fetal scalp. A second plate electrode on the mother's skin provides a reference, and electronic circuitry can pick up the electrical signals of the baby's heart. This ECG signal is somewhat more stable than that from an ultrasonic transducer, but it requires that the fetus be in the head-down position and that dilation is sufficient to allow placement. The electrical ECG signal can also provide certain information not available with the ultrasound signal. The ECG signal is used to display a numerical value and tone, as well as record on a chart, just as with the ultrasound signal.

Another function of the fetal monitor is to measure the relative strength of uterine contractions. This is important since the fetus is most likely to be distressed during contractions. The uterine contractions are measured by a transducer disk (called a TOCO transducer) that presses a central tab against the mother's abdomen. During a contraction, the muscles in the uterus become much harder, and the disk is pushed back into the transducer. These variations are measured and processed to produce a graph of relative contraction strength, which is displayed as a numerical value on the monitor, and also graphed on the same chart as the fetal heart rate. This allows a good correlation to be made between contractions and any possible fetal distress.

Some fetal monitors also measure maternal blood oxygen saturation (SpO_2), as low oxygen levels can be very harmful to the fetus.

Another option for a monitor is the ability to handle twins; this simply requires a second, independent ultrasound portion, one for each fetus. Graphs are both on the same chart.

Some fetal monitoring systems have telemetry, with which the transducers are plugged into a small, battery-powered transmitter. The transmitter sends the signals back to the monitor; this has the benefit of allowing the mother to walk around.

Application

Ultrasound gel is applied to the ultrasound transducer, which is positioned to give the best fetal heart signal; if a scalp electrode is used instead of ultrasound, the electrode lead is inserted via the mother's vagina and attached to the fetal scalp. In either method, once a suitable signal is obtained, the transducer and cables are secured. The TOCO transducer is then applied and secured, and electronically zeroed while between contractions. If the unit has SpO_2 capability, this transducer is usually placed on the mother's finger. All signals are checked and adjusted as necessary, and the chart recorder is started.

Also known as

(None).

Related devices

Pulse oximeters, ECG monitors, Doppler units, stethoscopes.

Where found

Maternity wards.

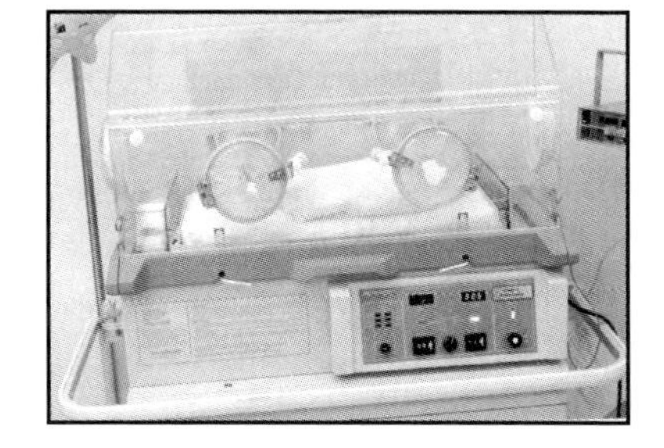

Infant Incubators

Overview

Neonates, especially if they are premature, need an environment in which factors such as temperature and humidity are controlled, and in which they are easily observable. Caregivers must be able to access the infant to change diapers, administer medication, feed, and simply provide physical contact, preferably without disturbing the controlled environment.

Function

Infant incubators are made up of a clear plastic chamber, electronic and mechanical systems to monitor and control the environment, and a stand to bring the chamber up to a comfortable working height. The stand is usually on lockable wheels and has storage compartments for diapers and other supplies.

The plastic chamber is designed to give maximum visibility from various angles, while providing reasonable insulation from exterior conditions and access to the infant. There is a larger hatch for moving the infant in and out of the incubator when necessary, and also several 'portholes' with flexible sleeves to allow caregivers to reach in to the baby, with minimal air leakage. The portholes are usually covered by a clear door when not in use. The outer wall of the chamber may be double layered to provide better insulation and also to control air flow; warm air may be directed between the outer layers and distributed into the interior at several points for more even heating. The chamber material must be strong enough for safety, but it must also be very clear for visibility, and allow penetration of ultraviolet rays for bilirubin therapy, if that is needed.

Temperature is the most critical environmental parameter to be controlled in an incubator. Neonates are often unstable in their internal temperature control; being small, and often having little body fat, their temperature can change quickly. It is important, though, for caregivers to be able to see as much of the infant's skin as possible, to watch for changes in color and texture which may indicate problems. Skin surface exposure is also important for bilirubin therapy.

Temperature may be controlled in one of two ways. Most commonly, a sensor measures air temperature and the reading is compared with the desired temperature, set by the caregiver. Then, the output of the internal heater is adjusted accordingly. This sensor may be built into the unit at some point in the air flow, or it may be a plug-in type with the actual temperature pickup suspended in the air above the infant.

The second method of temperature control involves placing a temperature sensor directly on the baby's skin. Again, this measurement is compared to the desired temperature set by the caregiver, and heat output is adjusted accordingly. To be effective, this method requires that the sensor be placed carefully; limb temperatures may vary considerably from torso temperature; the sensor must not be covered, and it must not interfere with treatments. Additionally, the adhesive used may irritate the baby's skin, and the sensor and wire may hinder removal of the baby for feeding, bathing, cuddling, or treatment outside the incubator. Given these considerations, incubators may not have skin temperature monitoring/control.

Temperature measurements are displayed on a front panel, which also shows the settings for desired temperature. There are alarms for over and under temperature, and usually a redundant high temperature alarm set somewhat above the primary high temperature alarm. Alarms are usually visual and audible.

Humidity control is usually simple, with a supply of sterile distilled water being placed in the air flow so that some of it can evaporate. Care must be taken to keep the water reservoir clean, as bacteria or fungi may grow in it, with potentially harmful effects. Air must be filtered to remove as much dust as possible, and these filters must be changed regularly.

Incubators normally have a port by which oxygen can be added to the interior air. Since oxygen levels must be high enough to be effective but not so high as to cause problems with infant retinal development, the oxygen concentration in the incubator should be monitored and alarmed.

It is important that air be well-circulated within the incubator to maintain even and controlled heat. A fan, usually located near the heating element, accomplishes this. Because of the many hard surfaces within the incubator chamber, and because infants have sensitive hearing, fan noise must be designed to be as low as possible. Also, since proper air flow is critical, most incubators have some kind of detector that will initiate an alarm if air flow is lower than normal.

Finally, since incubator function is critical, most have a power failure alarm, so that staff can take appropriate measures if power is interrupted.

Application

When the unit has warmed to the desired temperature, the infant is placed in the incubator in a safe and comfortable position, usually with only a diaper on. Oxygen is adjusted if needed (and an oxygen analyzer is used to check concentration levels) and humidity checked. If the unit is to be used in the infant skin temperature control mode, a sensor is placed on the baby's skin and connected to the appropriate socket. If ultraviolet therapy lights are to be used, they are placed at the appropriate height and checked for output. Ports and doors are closed to allow conditions to stabilize.

Also known as

Incubators, care-ettes.

Related devices

Oxygen analyzers, infant resuscitators, bilirubin therapy units.

Where found

Maternity wards, nurseries, pediatric wards.

Infant Resuscitators

Overview

The immediate postpartum period is critical for infants at risk, and effective aids to intervention must be directly available and easily applied. Infant resuscitators were developed to meet this need.

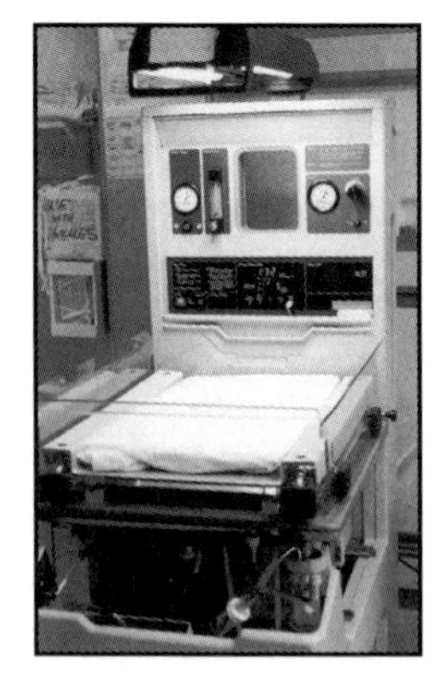

Function

Infant resuscitators consist of various components integrated into a single unit. There is a bed surface for the baby at a comfortable working height; this area usually has side walls, generally made of a clear plastic material for increased ease of observation, and to keep the baby in place. The side walls can be lowered or swung down for access, and are often marked in centimeters for approximate measurements of the infant. The bed surface usually has a chamber underneath to hold x-ray film cassettes. This means that the surface must be made of materials that are transparent to x-rays.

An overhead module provides bright lighting and radiant heating. The heating can often be controlled by a sensor on the baby's skin, so that a relatively constant body temperature is maintained even though most or all of the baby's body is uncovered. The overhead module is on a pivot so it can be swung out of the way when x-ray or other apparatus is being used.

The system has both oxygen and suction available to be used on the baby if needed; they are located so that they are out of the way when not needed, but readily accessible when they are required. There are indicators associated with these functions to show pressures and flow rates.

As part of the evaluation of the infant's condition, an Apgar test is usually performed: a set of observations made at specific intervals. A timer with visual and audible timing signals is often built into the resuscitator.

Application

The infant is placed on the resuscitator mattress and care is administered as required.

Also known as

Neonatal intensive care unit, NIC, Kreiselman (after an early manufacturer).

Related devices

Infant incubators, Apgar timers, gas regulators, aspirators, mobile C-arm x-ray units.

Where found

Labor/delivery areas.

Infant Scales

Overview

It is vital to keep track of the weight of newborns in order to monitor their progress. Infant scales are basically a simple scale but with a few important differences.

Function

Because babies might be moving vigorously, the scale must have sides to keep the baby in place. The scale, whether mechanical or electronic, must have some capacity to even out variations in readings caused by movements. There must be a 'tare' provision to allow for subtraction of the weight of diapers or blankets, and ideally there should be a 'hold' function to lock in a measurement when it is stable so it can be recorded later. Electronic scales usually have a switch to change between grams and pounds/ounces, and a large digital read-out.

Application

The infant is placed safely on the scale's weighing compartment. When a stable measurement is obtained, the value is recorded.

Also known as

(None).

Related devices

(None).

Where found

Nurseries, pediatric wards, emergency rooms, outpatient clinics.

Nitrous Oxide Units

Overview

Nitrous oxide ('laughing gas') is sometimes administered to patients who are experiencing relatively short-term pain, such as during minor surgery, dental procedures, or childbirth.

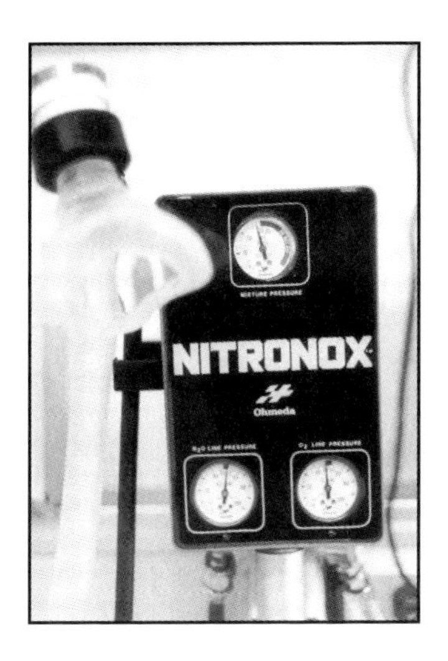

Function

Devices to provide safe and effective doses are generally simple, consisting of a source (either a central supply with a wall outlet, or a portable high-pressure tank), a regulator and gauge, a mixer to add air or oxygen to the nitrous oxide gas, a flow meter, a hose to carry the mixture to the patient, a mask to fit over the patient's nose and mouth, and a valve to open the line and deliver a dose to the patient. The valve may be manually operated, by either the patient or a caregiver, or it may be a 'demand' type that opens every time the patient breathes in.

Safety mechanisms include an alarm that sounds if either the nitrous oxide or air/oxygen pressures fall too low, and over-pressure valves to guard against failures that might deliver excessively high pressures to the patient.

Application

When required, a mask is placed over the patient's nose and mouth. Oxygen and nitrous pressures are checked, and the main valve is opened. Depending on the design, the patient may receive a dose continuously, at each breath, or when a button is pressed. Patients must be monitored carefully when using nitrous oxide to avoid over-exposure and to check for possible side effects.

Also known as

Nitronox (after a common model), nitrous.

Related devices

Anaesthetic machines, gas regulators.

Where found

Labor/delivery areas, outpatient clinics, dental clinics.

Oxygen Analyzers

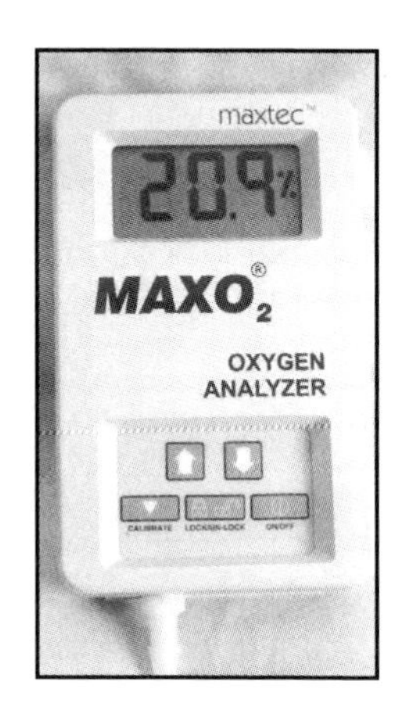

Overview

Whenever the needs of the patient for oxgen (O_2) are different than normal, it is necessary to determine the percentage of oxygen being delivered. This is particularly true for premature infants, as either too much or too little oxygen can be detrimental.

Function

Oxygen analyzers measure and display O_2 levels. They may have alarms that can be set for both high and low levels, and, as they are often battery-powered, there is usually a low-battery indicator.

Most oxygen analyzers work by directing a small amount of the gas being delivered to a patient onto a cell. This cell uses the oxygen present in a chemical reaction that varies with O_2 concentration; the reaction produces a voltage, which can then be processed and measured to determine the specific O_2 percentage.

Because the chemical reaction varies with factors such as temperature and the age of the cell, the system must be calibrated frequently to ensure correct readings. Calibration consists of exposing the cell to 100% oxygen and adjusting the display to match, and then returning the cell to room air, which is very consistently 21% oxygen. If the display shows 21% after settling, the system is ready for use; if not, it must be recalibrated, and if correct calibration is not possible, the cell must be replaced.

Oxygen enters the cell through a thin membrane, which must be clean and undamaged for proper functioning. Since the absorption of oxygen by the patient can be affected by many variables, it is often valuable to monitor blood oxygen levels (SpO_2) while oxygen is being administered.

Application

The unit is calibrated, and the sensor is placed in a location where it is exposed to the air to be monitored.

Also known as

O_2 analyzers, O_2/oxygen monitors.

Related devices

Oxygen concentrators, gas flow regulators, infant incubators, SpO_2 monitors.

Where found

Nurseries, pediatric wards, intensive care areas, operating rooms.

6. Operating Room

Because of the specialized nature of surgery and because patients are typically in the area for a shorter time than most areas of the hospital, operating rooms have a somewhat narrower range of devices than areas such as ICU or emergency departments.

Devices typically found in operating rooms, in addition to the equipment listed in this section, include: aspirators, capnographs, electronic probe thermometers, gas regulators, intravenous pumps, invasive pressure monitors, non-invasive pressure monitors, physiological monitors, point of care blood analysis systems, stethoscopes, syringe pumps, defibrillators, lithotriptors, and c-arm units.

Anesthetic Machines

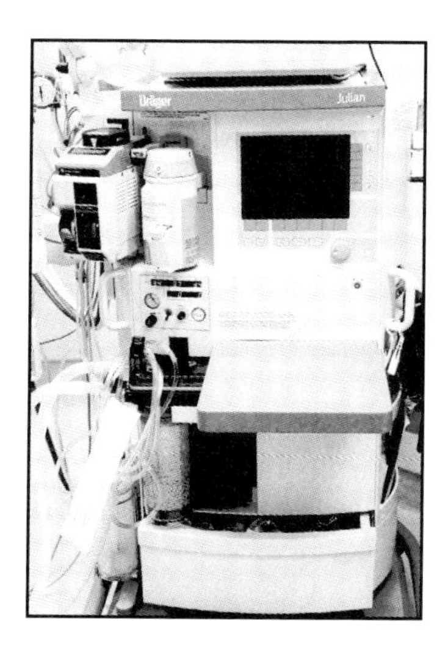

Overview

During surgery, there are a number of important considerations for the well-being of the patient, many of which are addressed by the attending anaesthesiologist using an anaesthetic machine.

Primary among these considerations is the elimination of the sensation of pain, which is accomplished by the administration of one or more gases that render the patient unconscious.

Function

The anesthetic machine provides a source for these gases, regulates their pressure and flow, mixes them with oxygen, adds humidification if required, and delivers the final mixture to the patient. Since the levels of the various components are critical, machines often have systems to measure and display percentages and flow rates, with associated alarms for high or low levels. This function may also be performed by an auxiliary unit, either stand-alone or integrated into a multi-parameter monitor.

Hypothermia is a common reaction during surgery, and so air provided to the patient may be heated; in this case, humidification is especially important to prevent drying of airway tissues.

Some surgical procedures require that the patient's muscles be immobilized to prevent unwanted contractions in response to either the physical trauma of surgery or to the electrical stimulation of electrosurgery units. Drugs to accomplish this are administered, but they paralyze breathing muscles as well, which means that the anesthetic machine must be able to provide artificial ventilation at a controlled rate and volume. This ventilation is delivered via an endotracheal tube, which in these situations also carries the anesthetic gases.

Anesthetic machines must also provide access to monitoring of vital signs such as ECG, blood pressure, temperature, blood oxygen saturation, and expired CO_2 levels. These functions may be built into the anesthetic machine itself, or the machine may simply be a platform on which an independent physiological monitor is mounted.

Application

Anesthetic machines are operated by physicians specialized in anesthesiology. They administer intravenous medications as required, then connect the patient to the anesthetic machine breathing circuit. After adjusting the appropriate parameters such as gas mixture ratios, gas flow rates, and patient breathing rate, the gas flow is initiated. Patient condition is closely monitored at all times to maintain adequate levels of anesthesia – enough but not too much.

Also known as

Anesthetic units.

Related devices

Physiological monitors, electrosurgery units, ventilators, nerve stimulators, muscle stimulators.

Where found

Operating rooms.

Electrosurgery Units

Overview

One of the problems encountered in surgery is bleeding from the many blood vessels of various sizes that are severed while incisions are being made. This is more prevalent in some types of tissue, and results in not only an increased loss of blood for the patient, but also decreased visibility for the surgeon and support staff.

Function

Electrosurgery units help to solve the problem of excess bleeding by using high-intensity electrical signals to perform cutting procedures. A pencil-type probe with a blade on the end (which comes in a variety of shapes for different circumstances) applies the electricity to the cutting point, while a large conductive grounding pad placed on a fleshy part of the patient provides a return circuit. Power levels are adjusted for different procedures. The electrical energy vaporizes the tissue immediately around blade, and there is enough energy spreading into nearby tissues to heat them to a point where many of the smaller blood vessels are sealed off (cauterized). Larger vessels must still be tied off, but much of the excess bleeding experienced with scalpel incisions is avoided.

In some circumstances, a very localized effect is needed, such as when performing tubal sterilizations. In such instances, a special electrode is used which passes current directly from one side of the probe to the other, passing through the tissue to be cauterized. This function is called 'bipolar' operation, as opposed to the more normal 'monopolar' operation using a surgical pencil and separate ground pad.

Experimentation has shown that different shapes of electrical signals produce different effects – more concentrated cutting, more general cauterization, or a blending of each. Some electrosurgery units utilize a highly-charged beam of gas (usually argon) to perform the cutting and cauterization, instead of a metal blade.

In some circumstances, only cauterization is needed and simpler devices, which use the same principles as full-function electrosurgery units but which only provide cauterization, are used. Simple cauterization generally requires less power and less precise control than cutting, so these units are usually smaller and less expensive.

Application

Electrosurgery units are operated by setting the desired power level and applying the blade to the target tissue, then activating the current by depressing a foot switch or using a hand-switch located on the operating tool. Specialized attachments such as loops and small forceps are used for laparoscopic surgery. Special attention must be paid in setup to ensure that the return electrode (ground pad) is placed on a fleshy area of the patient, well away from the surgical site. Because of the nature of the high-frequency energy involved, power can flow out of the patient's body through alternate pathways if the grounding pad is improperly placed or if it comes loose. These alternate pathways can cause burns to patients and/or staff, so alarms circuits are used which measure the degree of contact of the ground pad, and prevent operation of the unit if they are not within certain limits. A value that is too low can mean that saline or blood has made a path from the incision site to the ground pad, which can result in extreme burns or even fires.

There is some concern that particles of smoke produced by electrosurgery units may contain active virus particles and/or toxins. To help alleviate this problem, some operating rooms are equipped with smoke evacuator units that are used in conjunction with electrosurgery units.

Also known as

ESUs, electrocautery units, cauteries, Bovies (after an early manufacturer).

Related devices

Surgical lasers.

Where found

Operating rooms, outpatient surgery clinics.

Laparoscopy Systems

Overview

Traditional abdominal surgery is traumatic for the patient. Large incisions are a direct shock to the system. They require a high degree of anesthetization, which is itself hard on body systems, and healing times and risks of infection are both significant. The large scars left are a cosmetic problem for some patients, as well.

Function

Laparoscopic surgery reduces these negative aspects of surgery by reducing the incision size from ten to thirty centimeters or more, to as little as one or two centimeters. This is made possible mainly by advances in fiber optic technology; optical cables provide a flexible pathway for illumination and for conveying images of the surgical target to the surgeon.

In order for the surgeon to be able to see the target, the abdomen must be filled with gas (insufflated). This requires a gas control system that regulates and measures flow rates, delivered volumes and (sometimes) inflation times. To provide a constant degree of inflation, the gas supply must be able to sense and control the intra-abdominal gas pressure.

Some simple procedures are done with a simple lens and eyepiece, which allows the surgeon to directly observe the internal structures and instruments. Most procedures, however, are visualized by an electronic camera pick-up attached to the fiber optic cables. The camera feeds an electronics module that conditions the signal and sends it to one or more high-resolution color video monitors, which the surgeon observes. This allows other personnel to see the progress of the operation for teaching purposes and enables more effective assistance. Images are often recorded for future examination.

Because color is sometimes critical in identifying structures and diseased tissues, both the illumination and display components of the system must have high color accuracy. Light sources must provide high-intensity illumination, which can be either manually or automatically controlled; light levels may interact with the video system, as well, to provide optimum values.

A channel in the access tubing allows insertion of the actual surgical instruments; these may be mechanical cutting or dissection devices, specially-designed electrosurgery probes, or attachments to break up kidney- or gallstones.

Sometimes one tube is used for observing the site while another provides access for surgical instruments. Suction must also be available, for removing excess fluids from the site and for aspirating excised tissues.

Application

The patient is sedated or anesthetized as necessary for the specific procedure, and then a small incision through the layers of the abdominal wall is made to allow insertion of the tube or tubes. The abdomen is insufflated, and the surgical site located and cleared as much as possible. The surgery is performed and the area is checked, then the gas is removed, followed by the tubes. The incision is then closed.

Also known as

Lap systems, lap-chole systems (an abbreviation for laparoscopic cholecystectomy – gall bladder removal), lap-gyne systems (for laparoscopic gynecologic procedures).

Related devices

ESUs, gas regulators, lithotriptors.

Where found

Operating rooms.

Muscle Stimulators

Overview

Some surgical procedures, such as those using electrosurgery units, can cause a patient's muscles to contract sharply. This can cause problems for both staff and patient, and a muscle-paralyzing agent is administered to prevent this from happening. Before surgery commences, the anesthetist must determine whether muscle paralysis has been effective. Also, patients must be monitored to ensure that excessive doses of the paralyzing agent are not given. Since the patient is unconscious at these times, a means for testing the depth of paralysis is required.

Function

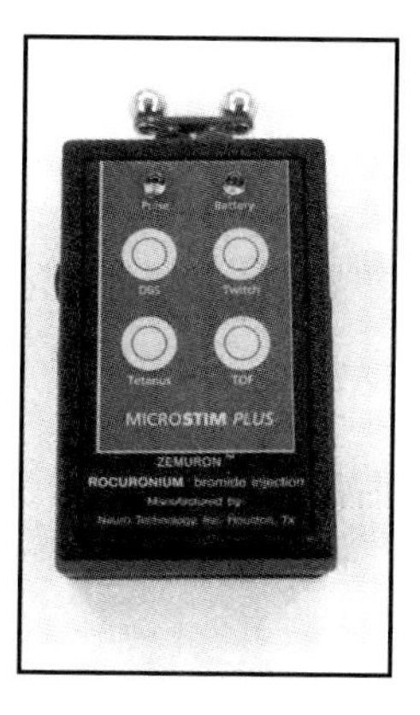

Muscle stimulators allow anesthetists to test for depth of muscle paralysis. Electrodes (which may be like ECG electrodes, or simple conductive rubber pads) deliver signals to the skin, where they are transmitted to underlying muscle. The intensity of signal required to produce a given response is a measure of the degree of paralysis. Controls on the unit allow for varying intensity, signal shape, and signal frequency.

Application

Electrodes are placed in a convenient location, usually on a limb near smaller muscle groups. When the patient is anesthetized, signals are applied to the electrodes, and the depth of muscle paralysis is determined. Tests are repeated frequently during surgery; if paralysis is too great, the flow of paralyzing agent is reduced, and vice versa.

Also known as

Stimulators.

Related devices

Electrosurgery units, Transcutaneous Electrical Nerve Stimulators (TENS).

Where found

Operating rooms.

Operating Room Lights

Overview

It is imperative that surgeons and support staff have a clear view of the operating field.

Function

Operating Room lights must supply illumination that is adjustable for both intensity and direction, evenly distributed so that all areas are illuminated evenly, and with a proper degree of 'whiteness' so that the colors of anatomical structures and tissues are natural.

OR lights are ceiling-mounted, with articulating arms to allow for a variety of positions. They generally have either a single, high-intensity bulb whose light is reflected from a large, carefully-shaped mirror, or a number of bulbs in a single housing, each with its own reflective area, to provide the even illumination required. Two or more lights may be mounted in the operating theater for increased illumination and placement flexibility.

Because intensity levels are high, a part of the design of OR lights is to reduce infrared heat, which could cause discomfort for staff and possible tissue drying for the patient. Mounting the light at a suitable distance from the OR table and using special coatings on glass surfaces minimizes heat effects.

Application

The lights are turned on and adjusted for optimal illumination.

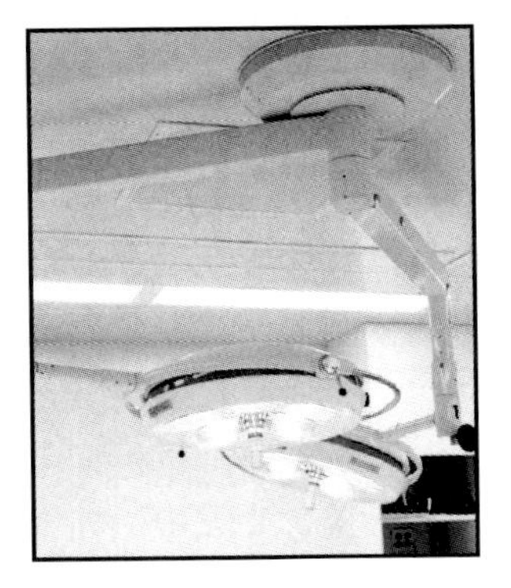

Also known as

Surgical lights.

Related devices

OR tables.

Where found

Operating rooms, some emergency rooms.

Operating Room Tables

Overview

During surgery, the patient must be in a stable, convenient, and adjustable position. There must be room for a number of people to have access, and there has to be provision for taking x-rays during the procedure.

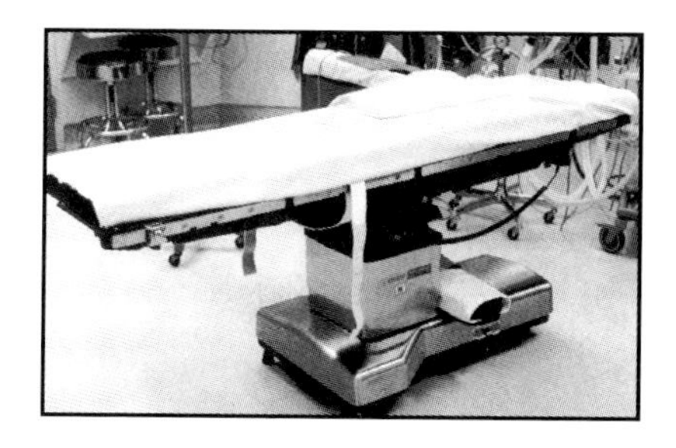

Function

OR tables are designed to move easily in three directions, as well as to be adjustable for head-to-foot and side-to-side angles. They are driven by either electric motors or hydraulics, though hydraulics are preferred because they are usually smoother and quieter, and safer (especially in regards to electrical shock hazards) when compared to electrically-operated versions. Movements are controlled by a footswitch or by hand controls, and there are locking mechanisms to hold the selected position solidly. An under-table chamber allows for insertion of x-ray film cassettes; some tables have a fluoroscope pick-up in the base, and some may have facilities for rapid repeat exposures of x-rays.

Application

Patient position is chosen to optimize access and comfort for the surgeon(s) and assistants. Certain surgical procedures and/or patient conditions may require the head of the table to be tilted up or down. The surgeon may perform adjustments, or they may be done by an assistant; position may be modified during the course of a procedure as well.

Also known as

OR tables, surgical tables.

Related devices

Room x-ray units, fluoroscopy units.

Where found

Operating rooms.

Operating Microscopes

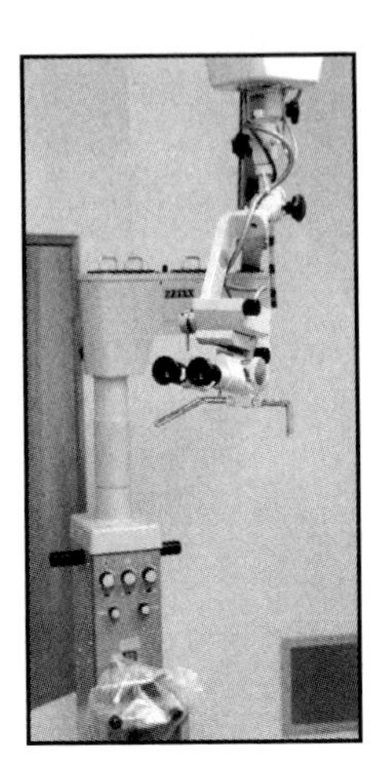

Overview

Certain surgical procedures involve very small structures, which require magnification to visualize and work on.

Function

Operating microscopes are one type of device used to provide this magnification. To be effective, they must meet a variety of design considerations. Primarily, they must provide an effective view of the surgical site. This requires high-quality optics in a binocular arrangement (to give a three-dimensional image), with variable levels of magnification and high-quality spot illumination.

For practicality, the surgeon and/or support personnel must be able to position the viewing head accurately, and it must remain stable during use. This requires a ceiling mount (in rooms that are dedicated to this type of procedure) or a heavy, wheeled floor mount. The head is attached to the mount with a relatively long arm that is multi-articulated to allow a range of positions. Vertical positioning and focusing are usually performed using a footswitch; other functions such as angle, horizontal position, and magnification may be powered or manual.

To accommodate the variety of positions needed, light is usually delivered to the viewing area via a flexible fiber-optic cable. This also has the advantage of reducing the amount of heat delivered to the site. Colored filters may be supplied, which help to highlight certain aspects of anatomy.

Operating microscopes may be equipped with dual viewing heads (looking through a single objective) for use by a student or assistant. They may also have a video camera so that the image seen by the surgeon can be displayed in a large video monitor, and possibly recorded for future reference.

Application

The surgeon views the surgical site through the microscope, adjusting position, focus, lighting, and magnification as necessary.

Also known as

OpMi, OR scopes, operating scopes/microscopes.

Related devices

OR lights, OR tables.

Where found

Operating rooms, specialized outpatient clinics, some emergency rooms.

Phacoemulsifiers

Overview

Cataract surgery is performed often, and the techniques of removing the clouded natural lens from the eye and replacing it with an artificial one are finely tuned. One aspect of the procedure, that of removing the old lens, is aided by a specialized device called a phacoemulsifier.

A primary objective in surgery is to minimize trauma, and to this end, incisions are kept as small as possible. This is especially true for ophthalmic surgery, but the ideal incision size does not allow for removal of the lens in one piece during cataract surgery.

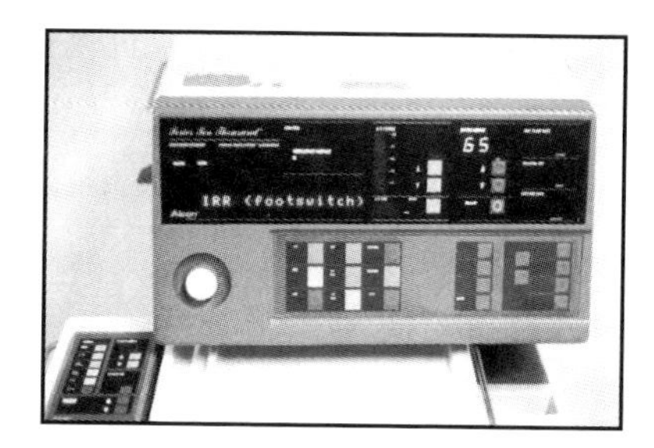

Function

The phacoemulsifier aids in this step by breaking the lens material into very small pieces (emulsifying it) and extracting the resulting product. A fine tip delivers very-high-frequency vibrations that break up the solid lens material; a parallel duct then applies suction to the area, removing the lens particles, while a second duct supplies irrigation fluid as required. Some units utilize a specially-tuned laser beam to break up the lens.

The unit requires controls for the various steps (these controls can usually be operated by a foot-pedal assembly, leaving the surgeon's hands free), as well as displays showing suction and infusion levels.

Application

Patients are sedated, but usually conscious during these procedures. Once the surgeon has established an appropriate incision, the phacoemulsifier tip is introduced into the eye and applied to the lens. The lens material is broken up and removed by the system. Some of the clear material within the eye may need to be replaced if it gets carried away along with lens particles. Once the entire old lens has been removed, the surgeon can begin installing the replacement plastic lens.

Also known as

Phaco machines.

Related devices

Operating microscopes, ophthalmic lasers.

Where found

Operating rooms, specialized outpatient surgery units or clinics.

Surgical Lasers

Overview

Certain surgical procedures require that tissues be vaporized in order to achieve the desired results, such as removal of skin abnormalities or cutting through tissue. Coagulation of tissues adjacent to the cutting site may also be desirable, or coagulation may be the goal.

Function

Surgical lasers utilize very pure and controlled light beams to heat tissue. Depending on the width and power of the beam, this heating may be extreme, vaporizing the tissue, or less so, in which case cauterization is produced. The laser beam may be controlled so that the depth of its effect is very precise, allowing removal of skin blemishes or tattoos.

A power supply and optical system develop the laser beam. Its characteristics are determined in part by the material used to generate the laser light. Some such materials are CO_2, Neodymium/Yttrium/Argon (Nd:YAG), and Erbium YAG. The laser beam is directed by a series of mirrors to the target area.

A low-powered laser beam, utilizing the same pathway as the treatment beam, may be used to point the system at the correct location before applying power. Control circuitry allows for adjustment of beam power and duration of treatment.

In order to properly visualize the target area, and to observe the effects of the laser beam, a microscope system may be used.

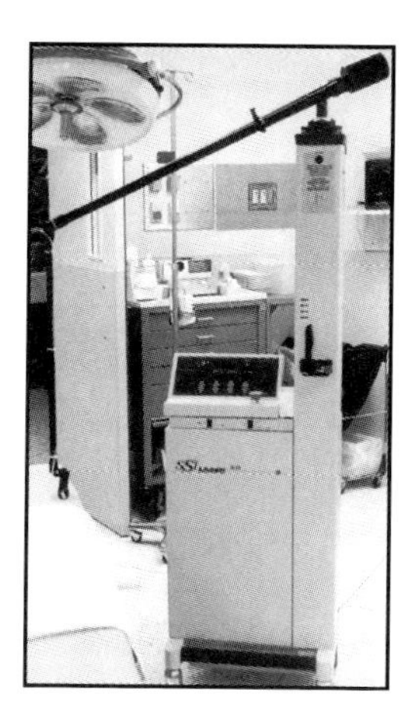

Application

The surgeon utilizes a hand-held probe to direct the laser beam, applying power by depressing a footswitch until the desired effect has been achieved.

Because laser beams may cause damage to the retina, either through direct exposure or by reflection, all personnel in the area must use proper eye protection, and adequate signs must be posted in the vicinity to warn that lasers are in use.

There is some concern that disease agents such as viruses may be present in the smoke produced by laser surgery. To guard against possible infections caused by these agents, a smoke evacuator system may be used to remove and filter out most of the smoke from the surgical site.

Also known as

(None).

Related devices

Ophthalmic lasers, slit lamps, operating microscopes, electrosurgery units.

Where found

Operating rooms, outpatient clinics, doctors' treatment rooms.

7. Outpatient Department

Outpatient departments may be in a hospital, often associated with Emergency Rooms, or they may be part of a medical clinic or doctor's office. In some smaller settings, they may be connected to the physiotherapy area.

Devices typically found in outpatient departments, in addition to the equipment listed in this section, include: electronic probe thermometers, tympanic thermometers, examination lamps, glucometers, non-invasive pressure monitors, oto/laryngo/ophthalmoscopes, sphygmomanometers, stethoscopes and defibrillators.

Cast Cutters

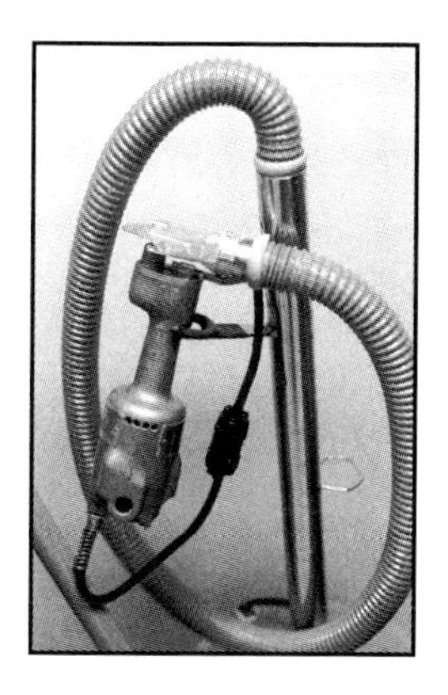

Overview

When it is time to change or remove a cast, the material (usually either plaster or fiberglass) must be cut. It is too hard and thick to cut with scissor-like devices, and because the patient's skin is very close underneath, conventional-type saws or knives are too dangerous.

Function

Cast cutter saws use a round, saw-toothed blade which is oscillated by the saw, rotating it a few degrees and then back very rapidly. The teeth are only moderately sharp, and, since skin is elastic, if the oscillating teeth should strike it, they merely push it back and forth without cutting. When the blade is applied to the hard, brittle cast material, it cuts through.

Application

Cutting a cast with a cast saw is simply a matter of turning the unit on and applying it to the part of the cast to be cut. Since it is noisy (and appears more dangerous than it is), and produces clouds of dust, patients need to be informed about the procedure in advance.

The dust from either type of cast material can be hazardous if inhaled, and is also very messy if spread around the room. Cast cutters often come with a vacuum attachment which sucks the dust from the cutting location before it can spread.

Also known as

Cast saws.

Related devices

(None).

Where found

Outpatient clinics, emergency rooms.

Cryosurgery Machines

Overview

Small skin abnormalities are often removed for clinical or esthetic reasons. Cutting the skin often results in considerable bleeding and opens the possibility of infection. Freezing the tissue concerned kills the cells, which are eventually are sloughed off without an open wound forming.

Function

While some situations requiring freezing may be handled by a simple swab soaked in liquid nitrogen pressed against the tissue, other circumstances require more control and longer application times. A cryosurgery machine uses the controlled expansion of a compressed gas to produce extreme cooling in a small metal tip, which can then be applied to the target tissue. Tips used are of various shapes and sizes, depending on the situation. Gases most commonly used are nitrous oxide and carbon dioxide; since both of these are potentially harmful (especially nitrous oxide), the units must be used in well-ventilated areas; preferable is to have the expended gas exhausted to the outside. The gas can be routed through the probe in such a way as to defrost the instrument at the end of the procedure.

Application

The operator depresses a control to release gas through the probe tip, continuing until the tip is well-cooled (indicated by a coating of frost and 'smoke' – wisps of freezing water vapor – around the tip). The tip is then applied to the target tissue, for a time determined by the size of the target and the depth of freezing desired. Additional gas may be used to keep the tip cold during longer applications. Defrosting is necessary to prevent the tip from sticking to the target tissue when treatment is finished.

Also known as

Cryo units.

Related devices

Electrosurgery units.

Where found

Outpatient clinics.

Electro-convulsive Therapy (ECT) Units

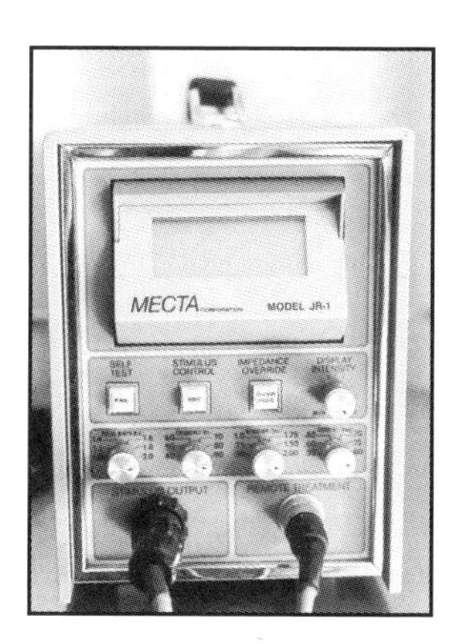

Overview

Certain psychiatric conditions seem to respond well to the administration of a controlled electrical signal to the brain.

Function

An Electro-convulsive Therapy (ECT) machine provides precisely shaped and timed signals (both parameters under control of the operator) via special electrodes placed on the patient's scalp.

Application

Since these electrical signals can cause powerful muscle contractions in other parts of the body, and for patient comfort, the patient is anesthetized or heavily sedated prior to the treatment, which is usually only a few minutes in duration. To be observed for possible side effects, the patient is usually placed on a physiological monitor during the procedure.

Also known as

Shock therapy units.

Related devices

EEG machines, physiological monitors.

Where found

Specialized outpatient clinics, post-anesthesia recovery areas, operating rooms, psychiatric special procedure rooms.

Endoscopy Systems

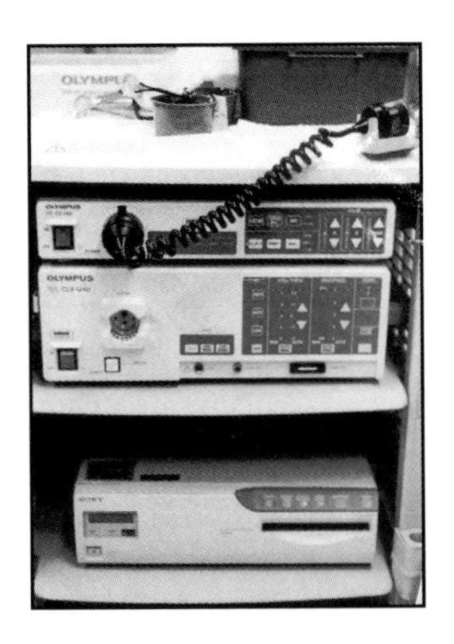

Overview

It is often necessary to examine the inner structures and surfaces of a patient's digestive or respiratory systems. Any means of performing minor surgical procedures (such as excisions of growths or cauterization of wounds) requires the ability to visualize the tissues in question.

Function

Endoscopy systems consist of several components: the central component is a tube, called an endoscope or scope. The endoscope has various parts, including: fiber optic light guides for illumination and observation; channels for supplying air to inflate (insufflate) the body cavities being examined and to remove such air after the procedure; another channel for irrigation and aspiration of the area; if surgical procedures are to be performed, a channel for devices used for these procedures; and, in some cases, a set of cables and bands that allow the operator to turn the tip in various directions.

The scope must be very smooth, both to prevent damage to tissues during insertion and removal, and to reduce the opportunity for infectious organisms to remain on the surface of the tubing. Scopes are generally marked with clearly visible rings that indicate the depth of insertion.

Another component of the system is the light source, which provides high-intensity, adjustable illumination for procedures. The light source may incorporate other components, such as an insufflator and irrigation aspiration pumps or simple electrosurgery unit circuitry, or these components may be separate. Insufflators allow adjustment of gas flow rates and pressures.

Finally, the system requires a means of displaying images. This may consist of a simple lens for viewing by a single operator, or, more commonly, a video camera, which produces video images for viewing on one or more video monitors. These images can also be recorded on videotape or other media.

Application

The patient is sedated or sometimes anesthetized before the procedure, as it can be physically and/or psychologically uncomfortable. The endoscope is introduced into the patient's system either through the mouth or the anus, depending on the part of the system to be examined. A lubricant aids in this process. The operator watches the images being produced to make sure the scope is inserted properly. Once the target area is reached, the cavity may be insufflated and irrigated if necessary, and examinations or surgical procedures are carried out. When the procedure is finished, as much air and irrigation fluid as possible is removed, and the scope is withdrawn.

Since large numbers of infectious organisms may be present during procedures, and may contaminate both the inside and the outside of the scope, special cleaning equipment is used to thoroughly sterilize the scopes between patients.

Also known as

Endo systems, scopes.

Related devices

Laparoscopic surgery systems, electrosurgery units.

Where found

Operating rooms, special outpatient clinics.

Ophthalmic Lasers

Overview

Various ophthalmic conditions such as astigmatism and myopia cause difficulties for the patient and must be resolved if possible. Corrective lenses provide adequate focusing improvement in most situations, but they may be unsuitable for some patients either for aesthetic reasons or for physical limitations or specific job requirements. Surgically reshaping either the cornea or the whole eye may produce satisfactory results.

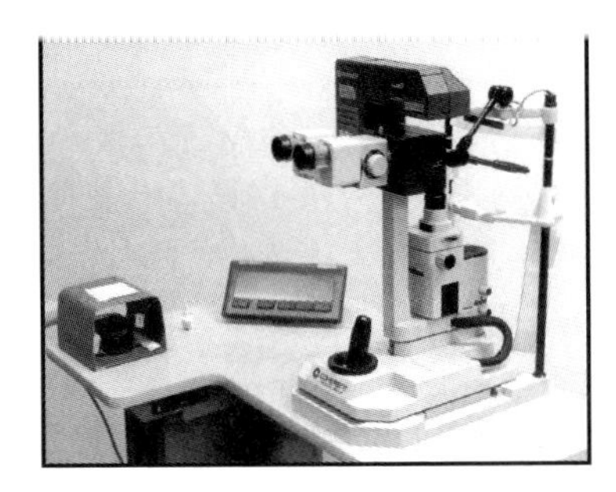

Function

Ophthalmic lasers allow surgeons to reshape the cornea to correct focusing problems. Specifically-selected light wavelengths, power levels, and application methods allow the cornea to be reshaped without incision, resulting in greatly shortened procedures and healing times. Precise control of the laser's action is vital to produce optically accurate curvatures and surface.

Application

The patient is positioned with chin and forehead resting on supports, to help maintain a precise position during treatment. The ophthalmologist views the target structures on the eye with the microscope portion of a slit lamp. After setting power and duration parameters, and aiming using a targeting beam, the treatment beam is triggered. Results are examined, and treatment continues until the final goal is attained.

Because laser beams may cause damage to the retina, either through direct exposure or by reflection, all personnel in the area must use proper eye protection, and adequate signs must be posted in the vicinity to warn that lasers are in use.

There is some concern that disease agents such as viruses may be present in the smoke produced by laser surgery. To guard against possible infections caused by these agents, a smoke evacuator system may be used to remove and filter out most of the smoke from the surgical site.

Also known as

(None).

Related devices

Surgical lasers, slit lamps.

Where found

Operating rooms, outpatient clinics.

Slit Lamps

Overview

The structure of the eye, both internal and external, must be clearly visible during certain examinations and surgical procedures. Since many of the structures involved are small, magnification is also important.

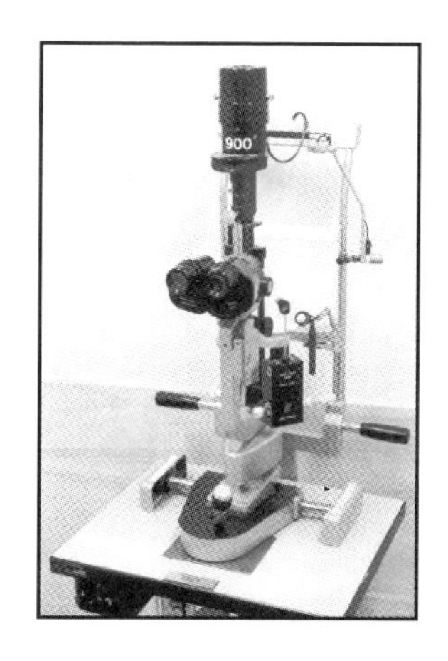

Function

A slit lamp is a combination light source and microscope. The light source can be focused to give a very narrow line, or slit, of high-intensity light, which aids in defining the curvatures and surfaces of ocular structures. A binocular microscope gives the operator a three-dimensional view of the areas being examined.

A camera may be attached to the microscope so that photographs can be taken.

Application

Fluorescein dye stains the outer surface of the eye, causing the eye to fluoresce under certain lighting conditions. Dye is sometimes applied to the patient's eye prior to a slit lamp examination in order to highlight various features of the eye. The patient places the chin in a cup-shaped device and rests the forehead against another curved support; this helps to reduce any motions that may interfere with examination or treatment. The operator then adjusts the shape and intensity of the light beam, and proceeds with the examination or treatment, viewing the illuminated eye through the microscope.

Also known as

(None).

Related devices

Operating microscopes, ophthalmic lasers.

Where found

Specialized clinics, outpatient departments, emergency rooms.

8. Physiotherapy

Physiotherapy departments may be connected with outpatient or cardiology/respiratory departments. They are very specialized, and the equipment found in physiotherapy is not usually found in other areas, unless they are transported there for use on a patient who cannot go to the physiotherapy department for treatment.

Some hospitals have a 'distributed' physiotherapy department, where much of the equipment is located in various areas of the hospital where it is used, rather than in one central place.

Continuous Passive Motion Systems

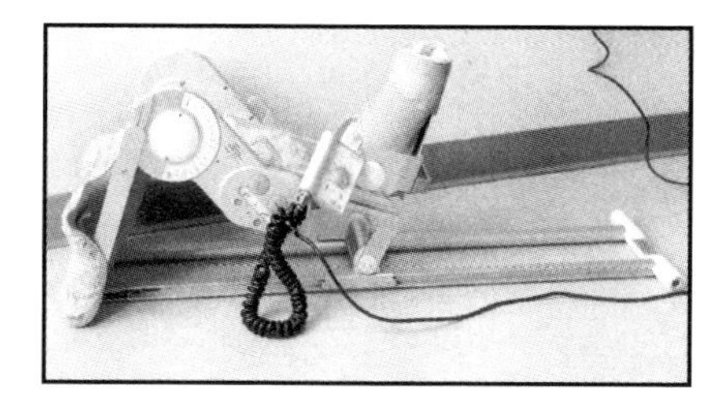

Overview

Patients are often unable to move their limbs enough for adequate rehabilitation. A mechanism that performs repetitive motion of the limb automatically is useful in promoting recovery.

Function

A continuous passive motion machine consists simply of a comfortable frame into which the patient's limb can be inserted and effectively immobilized, with mechanical hinges corresponding to the joints involved. A motor and gear or belt assembly then moves the frame back and forth on a track, flexing and extending the limb. The degree of motion is adjustable to suit each patient, as is the rate of operation. Systems often include a timer to signal the end of the selected treatment time.

Application

The patient is moved to a comfortable position that slows the range of motion developed by the continuous passive motion device. The target limb is placed in the frame, which is lined with a soft material such as sheepskin, and strapped in place. The frame has to be on a stable surface, though not necessarily smooth or horizontal. After determining the most effective range of motion and rate of action and setting the controls, the operator starts the machine, which can then function for a pre-determined time.

Also known as

CPM units or machines.

Related devices

TENS, sequential compression units.

Where found

Physiotherapy departments or clinics, patient bedsides where physiotherapy is administered.

Interferential Therapy Units

Overview

Pain reduction is a goal in the treatment of many medical conditions. Pain may not respond well to pain-killing medications, and even when it does, the medications used may have undesirable side effects, especially at high doses. Non-chemical means of alleviating pain can be very valuable.

Reduction of swelling (edema) in diseased or surgical sites is also important to promote healing.

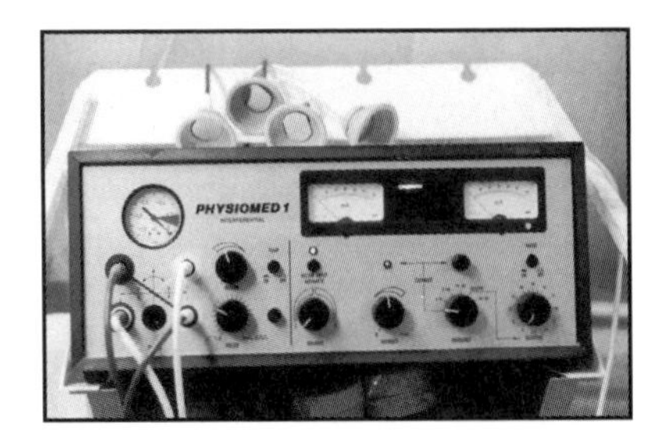

Function

In a manner similar to that of trans-cutaneous electrical nerve stimulator (TENS) units, interferential therapy units apply electrical signals to specific areas of the patient's body. These signals can reduce pain by either disrupting the normal pain nerve transmissions in the body, preventing them from reaching the brain to cause a sensation of pain, or by causing the body to produce endorphins, which are natural painkillers. The exact mechanism of operation is under debate.

Interferential therapy units use two pairs of electrodes, and pass signals in various patterns between the pairs so that an interference pattern of high and low intensity levels is produced in the tissues between and around the electrodes. This pattern apparently helps to reduce pain as well as reducing edema.

Some interferential units incorporate an alternating vacuum system, in which large suction cups fit over the electrodes. Vacuum is applied alternately to the suction cups, producing a massaging action. Units with this option are considerably larger than those without, since a vacuum pump and switching apparatus are required.

Application

Electrodes (and suction cups, if used) are applied to the patient's skin in a criss-cross arrangement over the target area. Power levels, rates, and patterns are set, as well as treatment time. An audible and/or visual alarm indicates the end of treatment, at which time the signals and vacuum are switched off.

Also known as

(None).

Related devices

Trans-cutaneous electrical nerve stimulator units, sequential compression units.

Where found

Physiotherapy departments and clinics.

Laser Therapy Units

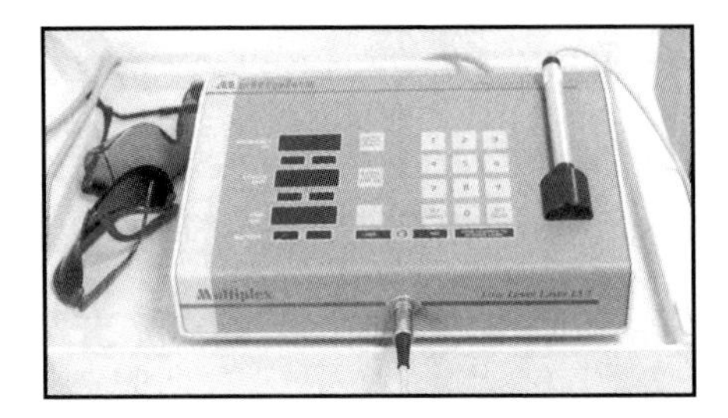

Overview

Reduction of pain and increased healing are two important goals in physiotherapy.

Function

Laser therapy units utilize a relatively low-power red and/or infrared laser beam to stimulate the skin. This stimulation can produce increased blood flow to the area, thus aiding healing. It has also been found to aid in reducing inflammation, as well as increasing electrolyte activity in target tissues, both of which can be beneficial to healing.

Laser therapy can also act in a way similar to TENS units or acupuncture by either blocking nerve signal transmission or stimulating endorphin (natural pain killer) production.

Units have controls for beam intensity, frequency, and treatment time.

Application

The patient is placed in a comfortable position with the target area exposed, and the operator sets treatment power, frequency, and duration. The laser beam is directed over the target region for the length of the procedure, possibly without further operator involvement.

Even though the laser beam used in these units is of relatively low power, it can still cause damage to eyes. The eyes of both patient and operator must be protected by goggles at all times, and the laser beam must never be directed at eyes.

Also known as

(None).

Related devices

TENS, interferential therapy, hot pack heaters, wax baths, ultrasound therapy, surgical lasers, ophthalmic lasers.

Where found

Physiotherapy departments.

Moist Heat Units

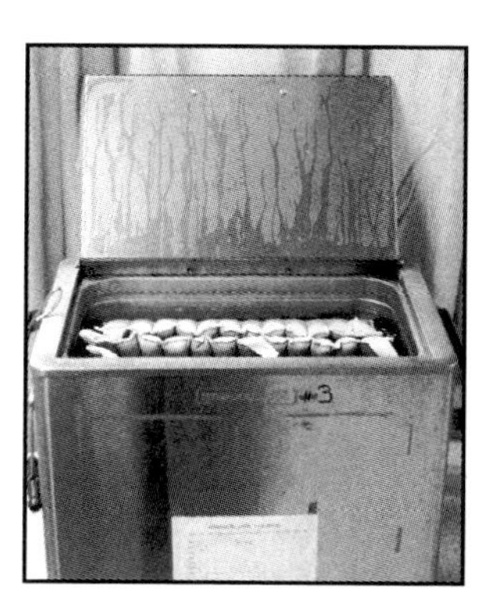

Overview

The application of heat can relieve some of the pain associated with tissue inflammation caused by injury or diseases such as arthritis. It can also promote healing by increasing blood flow to that area.

Function

Moist heat units are simply water heaters that store hot packs, which are cloth-covered bags that hold heated water and can be molded to the shape of the body part to which they are being applied. A thermostat maintains a constant, therapeutic water temperature in the water tank.

Application

Hot packs are placed against the skin of the patient to provide maximum heating of the target tissue. Towels or other insulating materials may be placed over the hot packs to help retain heat. After a pre-determined treatment time, the packs are either replaced or removed.

To avoid the possibility of contamination, water in the heater unit must be changed on a regular basis and tested for organism growth.

Also known as

Hydrocollators (after the model name of an early version), hot pack heaters.

Related devices

Wax baths, ultrasound therapy machines.

Where found

Physiotherapy departments and clinics.

Percussors

Overview

Various medical conditions (such as cystic fibrosis) can result in an excessive build-up of mucus in the patient's lungs. It is vital that these build-ups be removed on a regular basis, otherwise pulmonary function will be seriously impaired.

Function

Percussors are simple mechanical vibrators that send low frequency, relatively high power shock waves ('thumps') into the patient's body. The shock waves help to loosen the mucus so that it can be drained or expelled from the lungs.

Application

The patient is positioned head-down on a tilted table or frame; the percussor is applied to their chest and slowly moved around. The patient may need to be turned to various angles in order for treatment to be most effective.

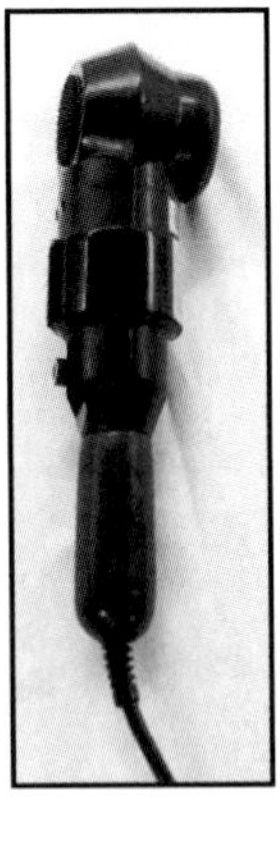

Also known as

(None).

Related devices

(None).

Where found

Physiotherapy departments or clinics, patient wards, patient's homes.

Sequential Compression Units

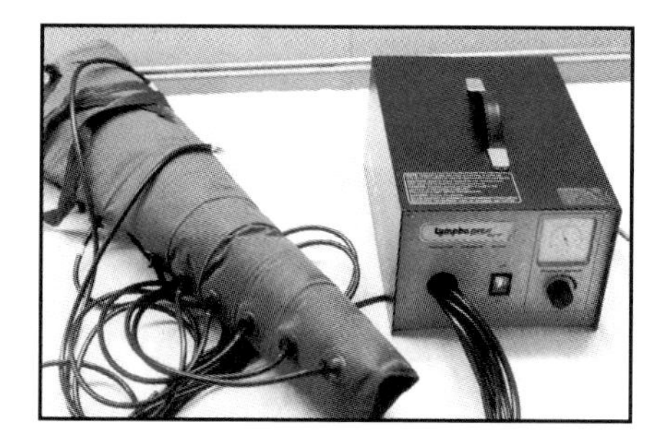

Overview

Fluids may collect in the limbs, particularly the legs, of patients with compromised circulation. This fluid build-up can lead to serious complications.

Function

A sequential compression unit consists of a sleeve that fits over the patient's limb. A series of air bladders in the sleeve can be inflated one after the other, from the distal part of the limb inwards, thus squeezing the limb with a peristaltic wave, pushing fluids toward the body. Some systems allow two such sleeves to be used at once, so that both legs can be treated simultaneously. An air pump feeds a controller unit, which directs air to the various bladders in proper sequence. Pressure and rate are adjustable, and a timer signals the end of treatment.

Application

Different sleeves are available for different limb sizes. Once the appropriate sleeve has been selected, it is pulled over the target limb and connected to the air hoses coming from the controller. After selecting pressure, rate and treatment time, the pump and timer are started. Treatment is passive on the part of the patient.

Also known as

(None).

Related devices

Continuous passive motion machines.

Where found

Physiotherapy departments or clinics, patient bedsides.

Transcutaneous Electrical Nerve Stimulators

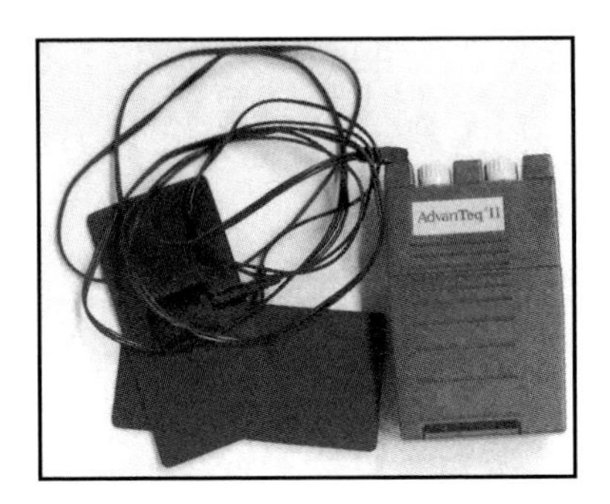

Overview

Pain reduction is a goal in the treatment of many medical conditions. Pain may not respond well to pain-killing medications, and even when it does, the medications used may have undesirable side effects, especially at high doses. Non-chemical means of alleviating pain can be very valuable.

Function

By applying controlled electrical signals to the skin at specific locations, a transcutaneous electrical nerve stimulator unit can produce significant pain relief. The exact mechanism of this relief is under debate, but it is thought that either the transcutaneous electrical nerve stimulator signals disrupt the normal pain nerve transmissions in the body, preventing them from reaching the brain and causing a sensation of pain, or that the electrical stimulation causes the body to produce endorphins, which are natural pain killers.

The unit producing the signals is generally quite small, and consists of a set of controls for signal amplitude, frequency, and shape, as well as visual indicators of treatment activity and connectors for electrode wires. The electrode wires carry the signals to electrodes (usually made of conductive rubber), which have been placed on the patient's skin.

Application

The operator may do some testing to determine the optimum placement of electrodes, by applying signals and checking for reactions from the patient. Sometimes electrodes are applied according to precise diagrams that specify locations for various sources of pain. In either case, the electrodes are affixed to the skin, treatment parameters are set, and the stimulation signals are turned on. Both the parameters and the electrode location may be adjusted during the course of the treatment, depending on patient response.

Since the electrical signals may interfere with pacemaker function, the use of transcutaneous electrical nerve stimulator with such patients is not advisable.

Pain relief may be an ongoing process, in which case patients often learn to use the transcutaneous electrical nerve stimulator unit on their own.

Also known as

Nerve stimulators, TENS units.

Related devices

Interferential therapy units, laser therapy units.

Where found

Physiotherapy departments or clinics, doctors' offices, patients' homes.

Ultraviolet Therapy Units

Overview

Exposing the skin to ultraviolet (UV) light can aid in the treatment of some skin conditions. Particular wavelengths of UV have been found to give the maximum benefit with minimal harmful effects. The timing of exposure is critical, as a balance must be reached between effective treatment and skin damage.

Function

UV therapy units are similar to commercial tanning booths, except that the special fluorescent UV lamps are designed to emit light most beneficial to medical conditions rather than to provide maximum tanning effect. Also, safety concerns are more carefully addressed, with timer circuits to prevent over-exposure.

Fluorescent tubes in the units are parallel to each other and are arranged in a circle, so that 360-degree, head-to-foot exposure is possible.

Because UV light output can vary with the age of the fluorescent tubes, operators must periodically check for light levels using a calibrated UV light meter, located a prescribed distance from the bulbs.

Application

The patient removes as much clothing as is required to provide adequate exposure, applies eye protection (as inadvertent UV exposure can be damaging to the retina), and then stands or lies in the treatment booth. The operator sets exposure time and turns on the lamps; at the end of the treatment period, the timer turns the lamps off and initiates an audible signal.

Also known as

UV booth, tanning booth.

Related devices

Bilirubin therapy systems.

Where found

Physiotherapy departments, special skin treatment units.

Ultrasound Therapy Units

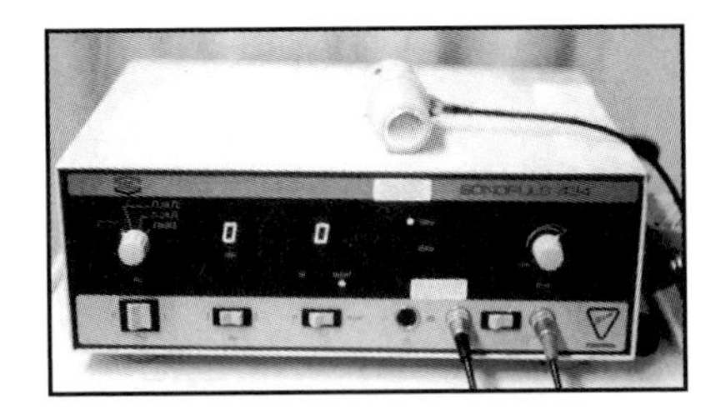

Overview

The application of heat can relieve some of the pain associated with tissue inflammation caused by injury or diseases such as arthritis. It can also promote healing by increasing blood flow to that area.

Function

High-frequency sound waves can cause localized heating in tissue. Ultrasound therapy units utilize this fact to provide such heating for therapeutic purposes. An electronic circuit develops high-frequency signals that are applied to a crystal, which in turn emits sound waves. By applying the crystal face (mounted in the head of a handle) to the patient's skin, the underlying tissue is heated. To ensure maximum transfer of sound energy from the crystal to the patient, a sound-conducting gel is placed between the two.

Devices often come with two or more treatment heads, each with different frequencies of operation. The different frequencies produce heating at different depths.

A means of controlling power output and a timer circuit complete the basic design of these devices. Some units have a visual indicator of contact quality, and possibly a means of reducing or cutting off power when coupling is inadequate.

Application

Ultrasound gel is applied to the face of the treatment head, which is then placed on the patient's skin before power is turned on. Either the operator or the patient moves the head to various positions to attain maximum heating benefit to the target area.

Power levels must be selected carefully to prevent burns, and care must be taken not to use the unit on areas with closely underlying bones for the same reason. Further, since an improperly coupled or open ultrasound head can get extremely hot (if the unit is not equipped with a contact quality safety monitor), power must never be applied unless the head is in proper contact with the patient.

Also known as

US therapy, ultrasound.

Related devices

Wax baths, hot pack heaters.

Where found

Physiotherapy departments or clinics.

Wax Baths

Overview

The application of heat can relieve some of the pain associated with tissue inflammation caused by injury or diseases such as arthritis. It can also promote healing by increasing blood flow to that area.

Function

A wax bath consists of a tub of paraffin wax that is kept just above its melting point by a thermostat and heater. The wax is formulated to have a relatively low melting point, so that it is liquid at non-scalding temperatures. A lid retains heat when the unit is not in use, and it is on a stand that places it at an accessible height.

Application

A patient's body part, such as a hand or an elbow, is dipped in the liquid wax and then removed, so that the wax coating solidifies. Repeated dipping produces many layers of wax, which retains heat for a considerable time. This heat is effective at penetrating into the tissue and effecting relief from inflammation.

Temperature of the wax bath must be monitored at all times to guard against overheating, even though most units have a double-thermostat control system.

Also known as

Paraffin bath.

Related devices

Ultrasound therapy, hot-pack heaters.

Where found

Physiotherapy departments or clinics.

9. Special Areas

This is a catch-all designation, and includes such areas as dialysis units, lithotripsy units, radiation therapy units, and EEG testing areas.

Devices typically found in these areas, in addition to the equipment listed in this section, include: aspirators, ECG machines, tympanic thermometers, examination lamps, gas regulators, intravenous pumps, non-invasive pressure monitors, physiological monitors, sphygmomanometers, stethoscopes, and defibrillators.

Electroencephalographs (EEGs)

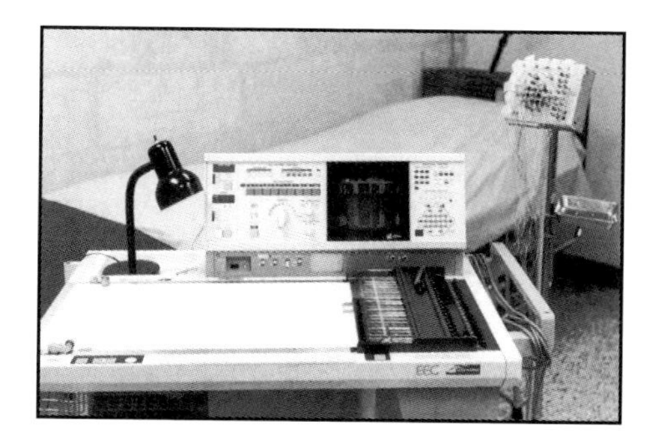

Overview

Electrical activity in the brain can give important information about its function and possible disease conditions.

Function

By taking signals from a large number of electrodes on specific points on the patient's scalp, amplifying and processing these signals, and then comparing and combining the various signals and analyzing the results, the electroencephalograph can present a recording of the signals that originate from various points within the patient's brain. The recording can be printed out on a chart, or recorded electronically for later examination. Most EEG machines incorporate a means of providing various stimuli to the patient, such as light flashes or sounds, which can then be related to changes in the electrical signals from the brain.

Application

Electrodes are applied to the patient's scalp according to a pre-defined pattern. Depending on the complexity of the analysis being performed, different numbers of electrodes are used. The operator checks connection quality and then proceeds with the test. The patient may simply lie quietly, or stimuli (such as sounds or light flashes) may be delivered. The timing of these stimuli is recorded so any changes in brain waves in response to the stimuli can be found.

EEG tests may be quite long, especially if relatively rare brain activity events are being studied. Portable EEG recorders have been developed that allow the patient to move around while measurements are being taken.

Also known as

EEG machines, brain wave machines.

Related devices

ECG machines.

Where found

Special areas, laboratories.

Hearing Testing Units

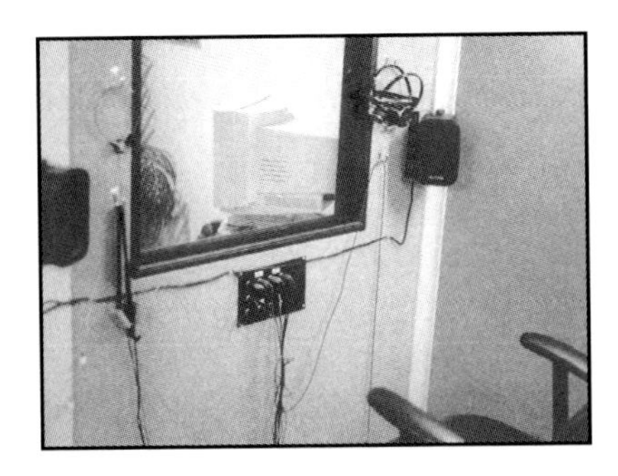

Overview

In order to diagnose hearing problems, or to properly design hearing aids, a very accurate measure of a patient's hearing ability must be obtained. This includes threshold perception levels at various frequencies for each ear.

Function

Since ambient noise can interfere with any hearing measurements, it must be reduced to insignificant levels. To obtain the most accurate measurements requires a booth with very sophisticated insulation designed into it, which may include lead sheet layers to help block outside sounds.

Test sounds, mostly pure tones at various frequencies, are generated by special circuitry and presented to the patient via high-fidelity headphones. The sound generator can vary both the frequency and the intensity of the tones, and can deliver them to each ear separately or in different degrees of combination. The test administrator can speak to the subject via a microphone and the headphones in order to give instructions and evaluate responses. A push-button or other mechanism allows the subject to signal the operator.

Application

The patient is seated comfortably in the hearing testing booth and given instructions regarding the testing procedure. Headphones are put on, and the tester withdraws from the booth. Various tones are sent to the headphones, in specific patterns. The frequency may be varied with a constant volume, or the volume may be varied with a constant frequency. Subjects indicate when they can no longer hear (or when they begin to hear) a particular signal either by hand gestures or by pressing a push-button. Results are recorded either manually or with a recording device.

Also known as

Sound booths.

Related devices

Otoscopes.

Where found

Special testing clinics.

Hemodialysis Units

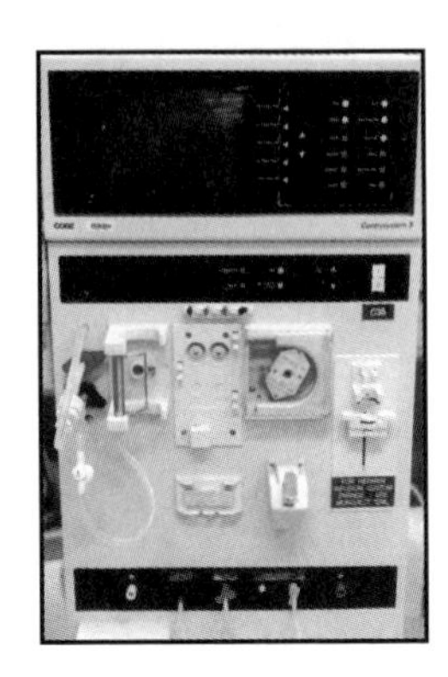

Overview

In cases where kidney function is extremely reduced or non-existent, or where peritoneal dialysis is not appropriate, patients may have their blood purified by hemodialysis.

Function

Blood is shunted from an artery in the patient's body and passed into a system that causes the blood to flow through a chamber where blood and a special dialysate solution are separated by a membrane film. The film allows some molecules to pass through but not others (and blocks blood cells) The dialysate solution is formulated to encourage toxic substances and excess salts and water to move from the blood, through the membrane, and into the solution. The system adjusts the temperature of the fluids involved so that the processed blood is close to body temperature. The cleansed blood is then passed back into a vein, usually adjacent to the artery from which it was removed.

Application

Once the arterial and venous ports have been established, the hemodialysis unit is filled with the various solutions required. Operation of the system is confirmed, and then blood flow is initiated. After the prescribed treatment time, the arterial flow is stopped and as much blood as possible is returned to the patient; then the venous line is closed.

In some patients, ports in the artery and vein can be semi-permanent so that they can be used for repeated dialysis procedures.

Also known as

Artificial kidney.

Related devices

Peritoneal dialysis units.

Where found

Specialized hemodialysis units or clinics; some hemodialysis systems are small enough that they can be used by appropriately-trained patients or family members in the patient's home.

Lithotriptors

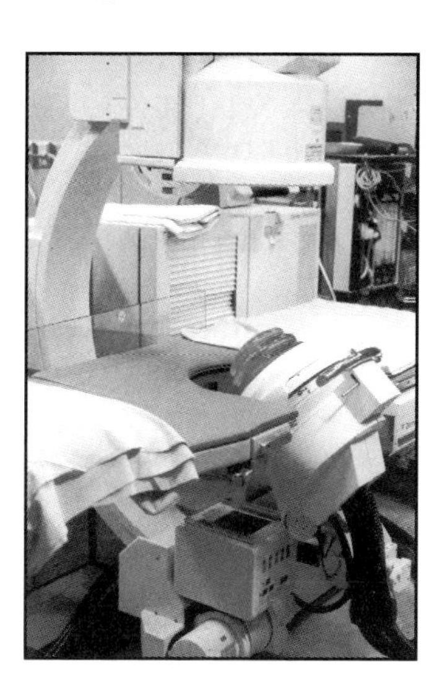

Overview

Kidney and gall stones (calculi) can be extremely painful, and often dangerous to the patient. Surgical removal is possible, but, like any surgery, carries a degree of risk, as well as an extended recovery time. If these stones can be broken up into small-enough fragments, they can be passed with minimal discomfort, thus avoiding the drawbacks of surgery.

Function

Lithotriptors (literally, stone breakers) use high-powered, high-frequency sound waves to shatter stones. Because the stones are much harder and more brittle than the surrounding tissues, the sound pulses affect them much more. When pulses are of sufficient magnitude and duration, they cause the stones to gradually break apart. To avoid soft tissue damage as much as possible, treatment times are generally extended. There are two general types of lithotriptors: direct contact and focused.

Direct-contact units consist of a special tip on a catheter. Circuitry passes a high-voltage pulse through the tip, causing a micro-burst of vaporized water. This burst in turn generates a sound shock wave, which begins to break the stone apart.

Focused systems utilize a number of sources that produce narrow sound beams. These beams can be closely focused on the stone from outside the body, and the convergence of the multiple beams is sufficient to break up the stone(s). These systems are further divided into two types, immersion and coupled. Immersion units have a bath-tub in which the patient sits, and sound waves are produced by transducers in the water. The water transmits the sound waves well, and they are transferred into the patient's body, where they are focused on the stone. Coupled units (pictured) use a rubber bladder in which the sound waves are produced. The bladder is pressed against the patient, with a coupling gel between to aid in sound transmission.

Application

Direct-contact lithotriptors require that the patient be sedated; a catheter is then introduced via the urethra, bladder, and ureter (for kidney stones) or through a small abdominal incision (for gallstones). The tip is positioned with the aid of a fluoroscope or an ultrasound machine. Pulses are initiated when the tip is in position, and continue, with regular monitoring to ensure correct position, until the stone is sufficiently reduced. This system requires a semi-invasive procedure, and some stones are inaccessible to the catheter.

Focused-beam lithotriptors are non-invasive, but tend to take longer to produce the same results as direct-contact units. This means that the patient must remain essentially immobile for the duration of treatment, which can be an hour or more. The patient is either immersed in a special water-filled tub, which houses the sound generators, or placed firmly on a rubber bladder (using a special gel to help couple the sound waves between the bladder and the patient's body). Sedation is required, partly to help the patient remain motionless and partly because the effects of the treatment on soft tissue adjacent to the stone can cause discomfort. The stone must be visualized and triangulated with a fluoroscope system; this location information is used by a computer system and the operators to keep the beams focused on the stone.

In all cases, patients are monitored to watch for possible adverse reactions to the treatment.

Some time (from hours to several days) after the treatment, stone fragments are passed by the patient. If treatment was ideal, the particles are small enough that little discomfort is experienced. Since kidney tissue is relatively delicate and highly vascularized, some bleeding into the urinary tract is likely to occur after treatment; this may also occur when stones are located in the ureter or bladder.

Also known as

Stone-breakers.

Related devices

Fluoroscopes, ultrasound machines, physiological monitors.

Where found

Special treatment units, special x-ray rooms, operating rooms.

Peritoneal Dialysis Units

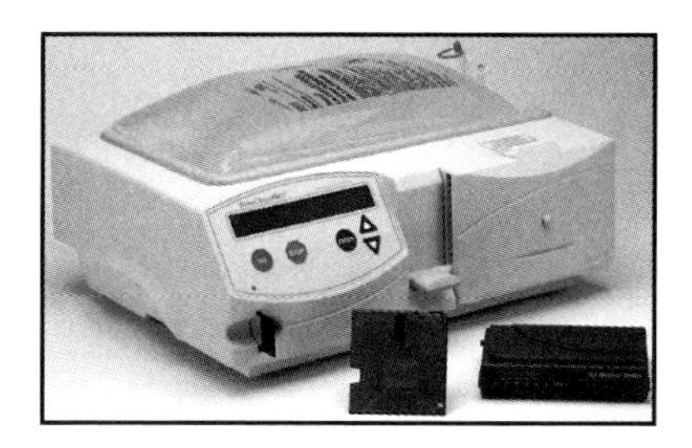

Overview

When a patient is experiencing severely reduced kidney function, toxins and excess salts and water normally removed by the kidneys build up in the body and must be removed to prevent serious harm or death.

Function

Peritoneal dialysis machines utilize two important principles. The first is that substances in solution tend to move from areas of higher concentration of that substance to areas of lower concentration; the second is that the inside of the abdomen (peritoneum) provides a large surface area that is well perfused. By filling part of the peritoneal space with saline solution, toxic materials in the blood will tend to move into the saline. The saline can then be drained, removing some of the toxic material as well.

Application

A catheter is inserted into the patient's abdominal cavity; this may be semi-permanent, and closed off when not in use. Saline solution is passed through the peritoneal dialysis unit, which warms it to near body temperature and controls the flow rate. The abdomen is filled with a volume determined by body size. The unit also functions as a timer, to signal when the saline should be drained. The process is repeated several times in order to maximize toxin removal. Volumes are recorded to ensure that no significant amount of saline is left in the abdomen. This technique has a number of drawbacks, including the trauma of catheter insertion, the possibility of infection, and the fact that some toxins may not be removed effectively, while some beneficial substances may be inadvertently removed. Some systems use smaller volumes of fluid for a longer time, with fewer repetitions, while others use larger volumes for shorter times, repeated a number of times. The smaller-volume process can allow the patient to be at least partially mobile during treatment, and is thus referred to as ambulatory peritoneal dialysis. The large-volume technique requires that the patient be essentially immobile for the duration of the treatment, and is referred to as cycling peritoneal dialysis; it can be done while the patient is sleeping.

Also known as

(None).

Related devices

Hemodialysis units, fluid warmers.

Where found

Specialized outpatient clinics (note: peritoneal dialysis is often done in the patient's home or even on vacations).

Radiation Therapy

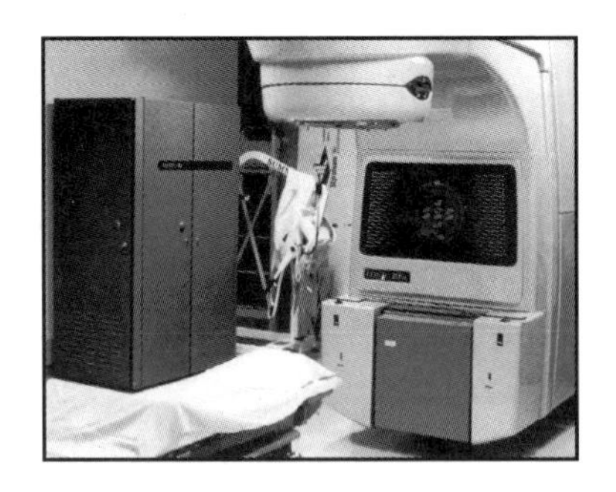

Overview

Some types of cancer cells can be killed by exposure to radiation; they tend to be more sensitive to radiation than normal cells because their rapid reproduction rate is disrupted by the radiation.

Function

By focusing the radiation from one or more sources on the area in which cancer cells are likely to be present, a radiation therapy unit allows the selective removal of cancer cells, while leaving most of the surrounding tissues relatively undamaged. A block of radioactive material (produced in high-technology nuclear facilities) is enclosed in a lead container, which has a window that can be opened to allow radiation to escape in a narrow beam.

Application

The patient is positioned so that the radiation beams can reach the target area with minimal penetration of healthy tissue. Because treatment times are quite long, the patient must be made as comfortable as possible while remaining motionless. When the patient has been positioned and non-target areas shielded, an operator remotely opens a window in the container holding the radioactive source, allowing the target area to be exposed to radiation. During the course of a treatment session, the angle of the radiation may be changed so that healthy tissue above and below the tumor site is irradiated less than the tumor.

Some tissues will selectively absorb particular radioactive materials, in which case (if there is a tumor in that particular tissue) relatively low concentrations of the material can be injected intravenously; the patient's body then concentrates the substance in the target tissue where it can deliver effective doses of radiation until it dissipates. An example of this is the thyroid gland's ability to concentrate iodine, including its radioactive isotopes. Obviously, this method doesn't require a radiation therapy machine, but many of the same precautions in handling the radioactive material are required.

Also known as

(None).

Related devices

PET scanners.

Where found

Special radiation therapy clinics.

10. X-ray Department

While most of the equipment found in an x-ray department is unique to that area, a few other machines may be located there.

Devices sometimes found in x-ray departments, in addition to the equipment listed in this section, include: aspirators, intravenous pumps, and stethoscopes.

C-arm Units

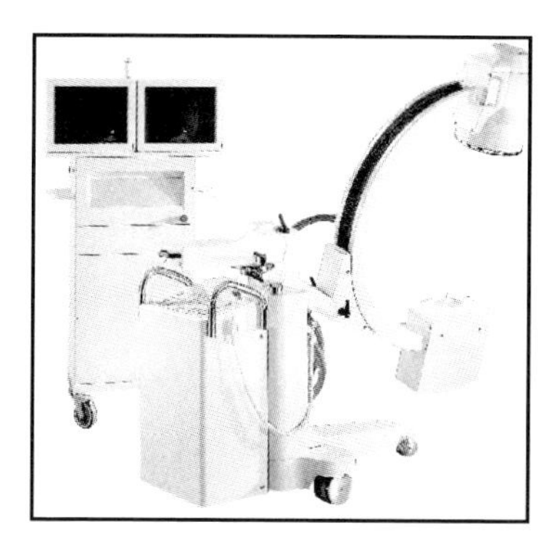

Overview

It is often necessary to obtain x-rays of patients from various angles without having to move them from their current position. This requires an x-ray machine that can be positioned so that its source is over the patient and its receiver is under the patient, and the whole assembly can be rotated to obtain the angled exposures needed. Some units are mobile, for use in situations where the patient cannot be moved to the x-ray department.

Function

Units were designed so that they could be moved, and then placed at the patient's location, with arm extensions that reach both over and under the patient. Since these extensions were in the shape of a 'c', the devices were called c-arm units. The upper extension houses the x-ray source and associated mechanisms, while the lower extension contains a chamber for film cassettes. The base of the unit allows for vertical positioning as well as rotation of the arm and extensions so that exposures at various angles can be obtained. In mobile c-arm units, the base is wheeled and has batteries to power not only the x-ray and positioning mechanisms, but also to propel the whole unit, as they are very heavy.

Application

The patient is prepared as much as possible and shielded if necessary. The c-arm is brought into place and adjusted much like a room x-ray unit, for position and exposure power and time. The operator (and any other personnel in the area) moves to a safe location, and exposures are made. The unit is then moved clear, and film cassettes taken for developing.

Also known as

(None).

Related devices

Room x-ray units.

Where found

X-ray departments, any area of the hospital where patients might be immobile but still require x-rays.

CT Scanners

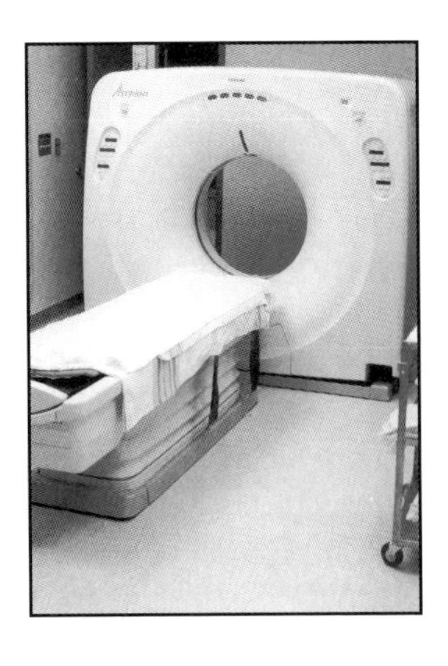

Overview

Conventional x-rays produce a 'shadow' image of the tissues through which the x-rays pass. Many disease conditions or abnormalities can be visualized better when a 'slice' view is available, so that structures are not obscured by tissues above or below, as may be the case with regular x-ray images.

Function

CT (Computerized Tomography) scanners pass a series of narrow beam x-rays through the target area of the body in a circular pattern, and then combine the information from each exposure with computers to develop a 'slice' image. With enough individual exposures and accurate computer analysis, very detailed images can be produced. Because the system uses narrow beams and a combination of many sub-images, each total image exposure can be relatively low, and the total radiation dose is acceptable. Newer devices use more sensitive x-ray receptors and faster computers to reduce the total radiation exposure and both the time required to acquire an image and the time to process the image and have it available for viewing. Images may be presented as an x-ray-like transparency, a printed picture, or on a video display; they can be stored as computer files for later retrieval and analysis. By taking a series of CT images, each adjacent to the next, a three-dimensional representation of structures can be developed.

Application

Patients are moved by a powered table surface into the scanning chamber, where exposures are made. Because exposure times can be somewhat prolonged, the patient must remain as still as possible; newer systems have reduced exposure times, but this is still a factor. The range of 'slices' that can be obtained is limited by the size of the scanning ring and the shape of the human body. Most systems have a ring not much larger than a human torso.

Also known as

CAT scanner, computerized axial tomography scanner.

Related devices

Room x-ray systems, MRI scanners, PET scanners.

Where found

Dedicated rooms in x-ray departments; some regions may have mobile units housed in a semi-trailer.

Diagnostic Ultrasound

Overview

Certain tissues cannot be visualized well with x-rays, and some situations (such as pregnancy) require that x-rays be avoided if at all possible, but visualization of internal structures is still needed.

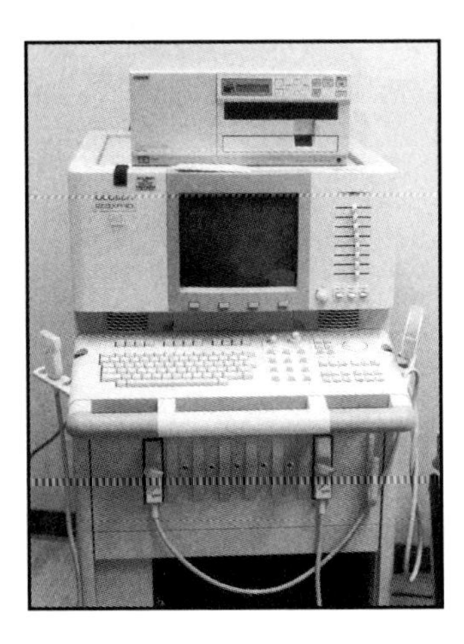

Function

Diagnostic ultrasound machines produce beams of ultrasonic sound waves, which can be directed into the patient's body. The sound waves are reflected by tissues of different densities, and by the boundary layers between different tissues. By picking up and processing the reflected signals, an image of the internal structures can be obtained. Various frequencies of ultrasound are used for different circumstances (such as the depth of the target organs), and probes are available that give wider or narrower beams. Visualization of a heart valve, for example, requires a narrower beam than one used for visualizing most or all of a developing fetus. The beam is scanned back and forth electronically to produce a full image.

Because of the nature of the wave generation and pick-up, raw ultrasound images represent a two-dimensional or 'slice' view of the tissue in question. While this may be sufficient for many applications, more sophisticated units can take the results of many such slices and combine them to give a three-dimensional view.

Devices may be general-purpose, or they may be specifically designed for particular applications, such as cardiology or maternity.

Application

The patient is positioned as required, and the ultrasound probe is applied to the skin. Because sound waves are partially blocked when they encounter a boundary such as between the probe surface and the skin, especially if there are any air gaps, a special gel that helps to couple the sound waves more efficiently is used. This also helps lubricate the skin to make positioning the probe easier. Particularly for fetal imaging, because images are taken from a variety of angles, large quantities of coupling gel are required. If the operators are nice, they keep the gel in a heated chamber so that it is warm when applied.

Also known as

US machines.

Related devices

X-ray equipment, doppler units, fetal monitors.

Where found

Special sections of x-ray departments, dedicated ultrasound departments; some devices are mobile and can be taken to the patient's bedside.

Dye Injectors

Overview

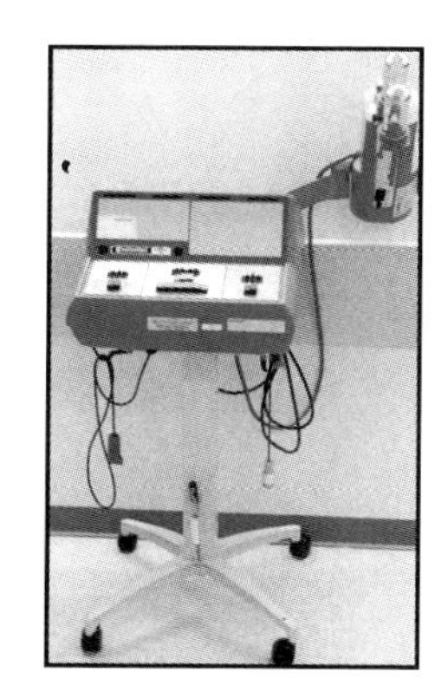

Some soft-tissue body structures, such as blood vessels, the urinary or digestive systems, and cerebrospinal fluid channels, can be better visualized on x-rays if a dye which is opaque to x-rays is injected into the structure in question before x-ray exposures are taken. In the case of the circulatory system, such injections must be rapid and precisely controlled in order to obtain useful and consistent results.

Function

Dye injectors consist of a syringe to hold the dye material and a mechanism to activate the syringe and inject the dye into the patient. Because high pressures are involved, the syringe and associated components must be designed to withstand these pressures. Also, since excessive length of tubing between the syringe and the patient will reduce the effectiveness of the injection, the syringe must be mounted on an articulating arm, which can position the components very close to the injection site.

Injection mechanisms must operate smoothly and have adequate power and control to deliver precision doses. The unit has controls to adjust volume and rate of injection, and a remote activating switch.

Application

An intravenous line is established at the required injection site, and the injector is loaded with dye and positioned close to the site and connected to the line. When the patient has been positioned for x-ray exposures, the operator moves to a shielded location and begins the injection. X-ray exposures are taken before, during, and after the injection, until the necessary information has been obtained.

Also known as

Angiography or angio injectors.

Related devices

Room x-ray equipment, fluoroscopy units.

Where found

X-ray departments.

Mammography Units

Overview

Early detection of breast cancer is critical in reducing harm from the disease, including mortality. X-ray imaging is one means of aiding such detection. Since cancer tissue is relatively close in x-ray density to normal breast tissue (as compared to bone, for example), and because tissue nodes that may be clinically significant are often small, means that can increase the effectiveness of imaging must be adopted. A reduction in the thickness of over- and under-lying tissue surrounding possible tumors is an important factor.

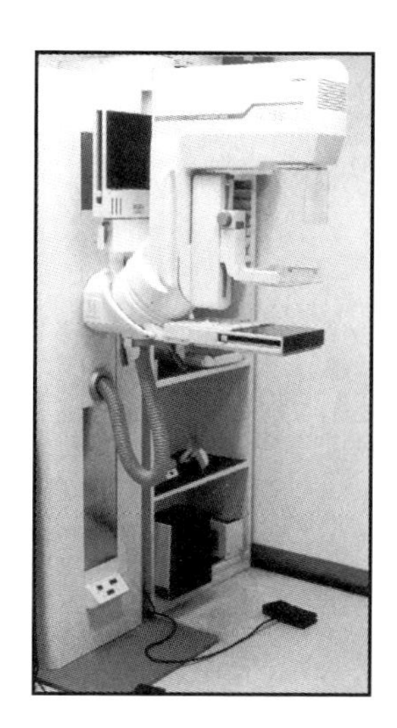

Function

Mammography units are similar to normal x-ray units, with x-ray characteristics optimized for breast tissue. They have attachments that compress the breast, horizontally and vertically, in turn. With the breast compressed, thus having a much thinner profile, x-ray film exposures are taken. If lumps have been detected by other means, these areas may be examined more closely.

Application

The operator sets power and exposure times for the procedure, and then adjusts the x-ray head to the patient. The patient, with the assistance of the operator, places one breast on a clear plastic plate. A motorized mechanism then lowers a second plastic plate onto the opposite surface of the breast until the tissue is adequately compressed. The operator then moves to the control area (shielded by lead glass and walls) and makes the exposure. The breast is then compressed along the other axis and exposures taken again. Patients are shielded appropriately to reduce x-ray exposure to non-target areas of the body.

As the compression required for good images is significant, this procedure can be quite uncomfortable for the patient.

Also known as

Mammo unit.

Related devices

Room x-ray units.

Where found

X-ray departments, special mammography units (including mobile ones).

Magnetic Resonance Imaging (MRI) Scanners

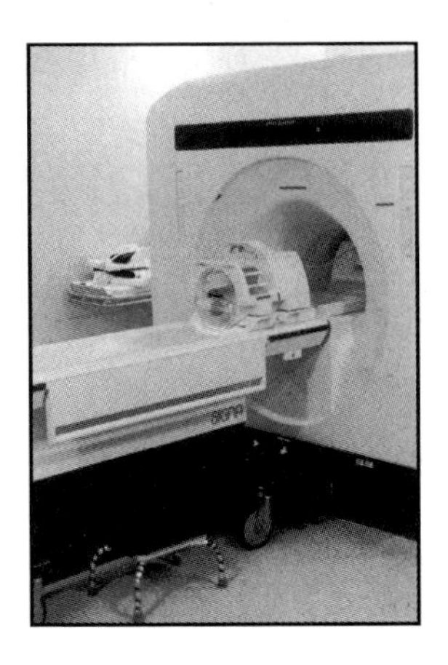

Overview

Since the human body is composed largely of water, a means of imaging the distribution of water in the body provides a picture of various structures giving a different kind of information than that provided by x-rays or other scanning technologies.

Function

Water molecules consist of two atoms of hydrogen and one of oxygen. When they are within a powerful magnetic field, radio waves of a precise frequency can interact with the hydrogen atoms and cause them to resonate and emit their own radio signals. These signals can then be detected and analyzed, and an image of hydrogen (and thus water) distribution within the body can be obtained. With sufficiently-advanced detection and analysis, very detailed images of body tissues can be obtained. These images can then be displayed on a video monitor and stored and/or printed out for later examination.

Application

Since extremely powerful magnetic fields are used in MRI procedures, all magnetically-active metallic objects must be removed from the patient. This means that patients with stainless steel implants or pacemakers may be excluded from such studies. The operation of pacemakers is also affected directly by magnetic fields, so that further restricts MRI use with these patients. Dental fillings and crowns are not usually affected.

The patient is positioned on an examination table, which can be moved around to facilitate imaging of different areas. The operator moves to a shielded location, and the magnetic field is initiated. To obtain the images desired, patient position is adjusted by the table, under both operator and program control. Typically two to six images are taken during a procedure, and since each image takes a few minutes for positioning and exposure, the whole procedure may take 15–45 minutes. This time may be longer for especially-detailed or complex studies. Patients must remain still during the actual imaging, but can shift to a limited degree between exposures.

A contrast medium may be injected into the patient for some studies, in order to improve image quality.

Also known as

MRI scan, magnetic resonance imaging.

Related devices

Room x-ray equipment, diagnostic ultrasound, PET scan.

Where found

X-ray departments, special clinics, mobile MRI units.

Proton Emission Tomography (PET) Scanners

Overview

It can be important to know which areas of the body are active metabolically in determining disease or injury conditions in a patient. Since specific compounds are involved in certain metabolic functions, a means of identifying the location and rate of uptake of these compounds will give information about metabolic activities in the tissue or organ in question.

Function

Various molecules utilized in metabolic activity within the body can be 'labeled' with radioactive elements that either replace the normal atoms of that same element within the molecule or attach themselves to the molecule in such a way that chemical activity is effectively unchanged. The resultant molecules are produced in a huge, high-tech machine called a cyclotron, and are referred to as 'radiopharmaceuticals'. These radioactive atoms are then carried through the body along with the whole molecules and are concentrated within organs or tissues that utilize these molecules and are currently active.

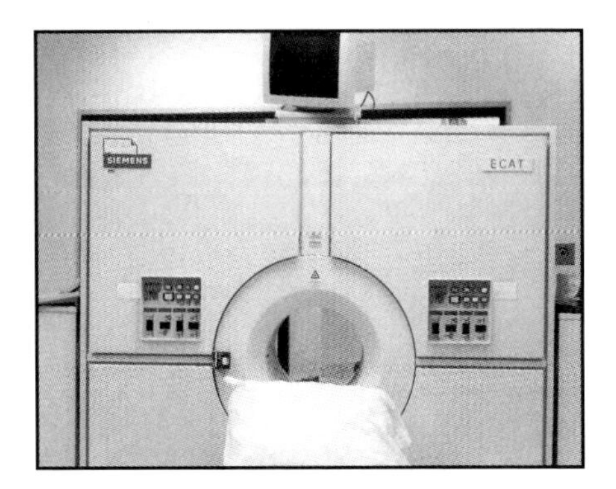

One example of such a molecule is fluorodeoxyglucose, which acts like normal glucose and is absorbed by any tissue in the body that is active metabolically. This is particularly useful in the brain, where tumors, injury sites, or areas involved in mental processes typically are more active than other nearby areas.

A second radiopharmaceutical is radioactive iodine, which is concentrated in the thyroid gland, especially in tumors or abnormally active regions. Conversely, iodine is less concentrated in abnormally inactive areas.

There are many other such radiopharmaceuticals, each of which is specific to certain types of metabolic activity in particular tissues or organs. They are used to help diagnose disease conditions in the target areas, or to perform research studies involving those areas.

Once the radiopharmaceutical has been concentrated in the target tissue, the actual PET scan can be performed. Radioactive atoms give off protons in the form of gamma-rays, which can be detected by a device called a gamma camera. By arranging as many as 180 of these gamma-ray detectors in a circular pattern around the target area, gamma-ray activity can be measured. Combining signals from the various detectors and processing with a computer system gives 'slice' images, which are colored artificially to show different degrees of metabolic activity in the image area.

Each color assigned to the image by the computer corresponds to a particular level of activity; this makes it easy to differentiate between active and inactive regions.

Some PET scan systems can take a series of adjacent slice images and combine them to give a three-dimensional view. Images are viewed on a video display and can also be stored or printed for later examination.

Application

The radiopharmaceutical chosen for the particular study is administered to the patient intravenously. After a pre-determined time, which varies with each type of examination, the patient lies on a moveable table, which slides them into the detector ring. Movements of the table are controlled by the operator and/or the computer system to position the patient in order to obtain the desired images. Images may be obtained in a series of slices to produce a three-D picture, and several exposures may be taken over time to observe changes in metabolism.

Radiation produced by the chemicals involved is relatively low-level, and the molecules either decay to harmless levels in a short time, or they are excreted by the body, or both. Since operators are potentially exposed to more radiation than individual patients, they must take adequate precautions to avoid over-exposure.

Also known as

PET scanner, PE scanner.

Related devices

CT scanner, MRI scanner, room x-ray equipment.

Where found

X-ray departments, specialized facilities, mobile units.

Room X-ray Units

Overview

X-rays of the appropriate power will penetrate tissues. If an appropriate detector is placed on the opposite side of the body from the source of the x-rays, a 'shadow' image is obtained which can give vital information about various structures within the body, since different body parts and tissues block the x-rays to a greater or lesser degree.

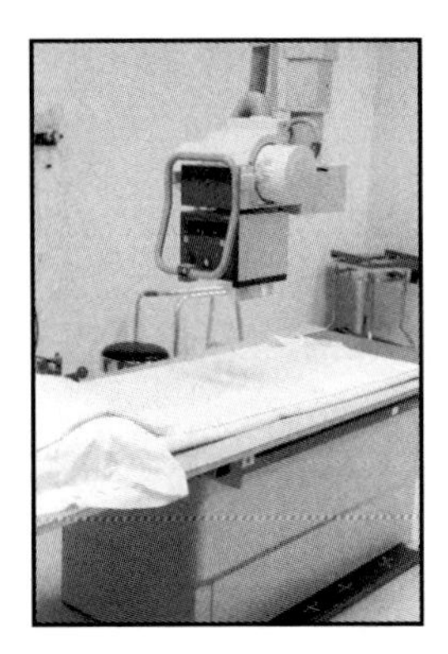

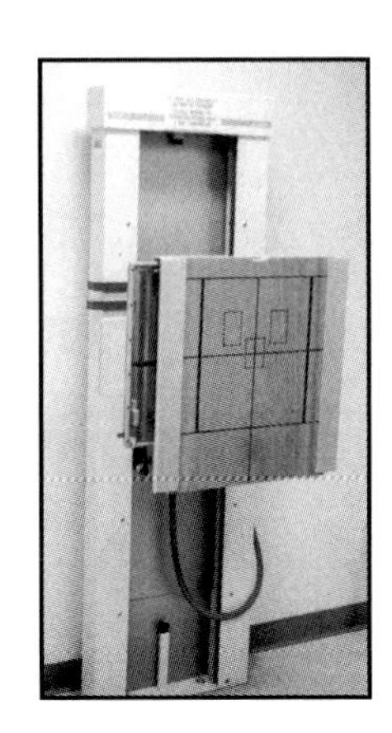

Function

By focusing a high-energy beam of electrons onto a spinning metal disc, the atoms of the metal can be excited to a point where they give off x-rays, an extremely high frequency form of electromagnetic radiation. Other examples of electromagnetic radiation are light rays, infrared or heat radiation, and radio signals.

The most common means of detecting and recording the images produced by an x-ray machine is silver-halide-based photographic film. This film is similar to everyday film used in cameras, but is generally much larger. It is optimized for x-ray exposures rather than light exposures, and it is very fine grained so that maximum detail can be seen in images.

Electronic components act to focus the x-ray beam at the plane of the film, and thus care must be taken to place the film and the x-ray source at the correct distance from one another.

Since the x-rays themselves cannot be seen, a visual system of aiming is used, with a light shining from the same point as the x-rays and making a shadow of cross-hairs. This part of the unit also has adjustable edges so that exposure can be limited to the specific target area.

Adjusting the power and time of the x-ray emission controls exposures. Power is measured in units called kilovolt-amps, or kVA. Different tissue densities and thicknesses, and varying requirements for detail and speed of exposure, determine these parameter settings.

Certain structures can be better seen on x-rays if a material that blocks the rays is placed into spaces within the body. Examples of this include a liquid form of barium which the patient may drink to help define parts of the upper digestive tract, or which may be given by enema when the lower parts of the digestive system are being investigated. Radiopaque dyes may be administered intravenously to show either blood vessels or the urinary system (after the dye is removed from the bloodstream by the kidneys).

Some x-ray units have mechanisms to automatically and rapidly change several film cassettes, in order to follow the progress of the barium or dye mentioned above as it moves through the patient's body. Others have a roll of film, as in a movie camera, to record a greater number of frames (though in a smaller format).

Excessive exposure to x-rays can be harmful, especially to reproductive organs, and so patients are normally draped with flexible lead aprons and/or sheets to limit exposure to the target area. Staff members use this equipment over long periods of time and so must take precautions to avoid over-exposure. Lead aprons are worn if they must be in the vicinity of the patient during exposures, otherwise they move behind a leaded shield for the time of exposure. Film badges are worn which can help evaluate the total dosage received over a period of time, and if levels reach certain thresholds, staff may be reassigned for a while.

Application

Patients are asked to remove any clothing from target areas of the body, and to replace it with a thin gown in order to minimize blocking of the x-rays. They are then placed in a position (either on a table or standing) that will allow the best view of the target, and distances and areas to be exposed are set. Since exposure times are often measured in relatively large fractions of a second, or sometimes even multiple seconds, patients must remain still during exposures. If the target area is in or near the chest, patients are asked to hold their breath for the exposure.

Barium is usually administered before the x-ray procedure starts, though it may be given while making exposures if, for example, swallowing actions are to be examined. Radiopaque dyes are usually injected immediately before x-rays are taken; pre-injection images may be taken for comparison purposes.

Also known as

Radiography units.

Related devices

Mobile c-arms, fluoroscopy units, dye injectors, CT scanners.

Where found

X-ray departments, mobile x-ray trailers.

Index